Zebrafish Model in Medical Research

Edited by Geonildo Rodrigo Disner

Published in London, United Kingdom

Zebrafish Model in Medical Research
http://dx.doi.org/10.5772/intechopen.1009050
Edited by Geonildo Rodrigo Disner

Contributors
Ali Reza Khansari, Ana Catarina Certal, Antonella Damiano, Anupam Srivastava, Carla Lima, Dayanand Hiremath, Geonildo Rodrigo Disner, Georgios Strimpakos, Guilherme Antunes, Huijie Yu, Hui Zhi, Inês Oliveira, Joana Monteiro, Juan Ramos, Kuo Zhang, Laura Sánchez, Laxmi Bhusal, Lluis Tort, Mariana Sampaio, Meijun Pang, Monica Lopes Ferreira, Priya Gupta, Sandra Martins, Thakur Rakesh Singh, Xiaoli Gao, Xiuyun Liu, Álvaro Arana

First published in London, United Kingdom, 2026 by IntechOpen
IntechOpen is the global imprint of INTECHOPEN LIMITED, registered in England and Wales, registration number: 11086078, 167-169 Great Portland Street, London, W1W 5PF, United Kingdom

For EU product safety concerns: IN TECH d.o.o., Prolaz Marije Krucifikse Kozulić 3, 51000 Rijeka, Croatia, info@intechopen.com or visit our website at intechopen.com.

British Library Cataloguing-in-Publication Data
A catalogue record for this book is available from the British Library

Zebrafish Model in Medical Research
Edited by Geonildo Rodrigo Disner
p. cm.
Print ISBN 978-1-83635-016-3
Online ISBN 978-1-83635-015-6
eBook (PDF) ISBN 978-1-83635-017-0

Meet the editor

Geonildo Rodrigo Disner is a Brazilian researcher and biologist holding a Ph.D. in Genetics with a focus on Environmental Toxicology and Mutagenesis. He is a former Postdoctoral Fellow in Toxinology and Immunoregulation (Zebrafish Platform) at the Butantan Institute and in Systems Toxicology at the Institute of Environmental Medicine, Karolinska Institutet, Sweden. Currently, he works as an associate researcher at the Zebrafish Platform at the Butantan Institute, coordinating the 'Microbioma' project, a research endeavor focused on the microbiome-gut-brain axis in zebrafish, supported by the Brazilian agency CNPq.

Contents

Preface

As an enthusiastic biologist and researcher who has spent years working with zebrafish, I have reached two convictions. The first is that the charm of this small fish is endless, remaining as captivating now as it was during the first encounters. The second is that the magnitude of possibilities offered by the zebrafish demands that the researcher remain in a state of constant study and learning.

This is arguably the book's primary purpose: to gather a body of relevant knowledge that reflects the major contributions of the zebrafish experimental model, specifically in medical research.

In the medical and biomedical field, new discoveries are made every year in both basic and applied science. Consequently, the zebrafish has been increasingly incorporated into research and development of products and services. The most popular applications in this area include drug screening, the study of genetic diseases and cancer, regenerative medicine, neuroscience and behavior, and environmental health.

In this book, the reader will find updates on the advantages and applications of zebrafish for medical science. It opens with an introductory reflection on the history of research with zebrafish, tracing a path that honors past achievements while pointing toward future perspectives. Subsequent chapters are organized to highlight specific applications, including two chapters focused on neuroscience and the treatment of neuropsychiatric diseases. These contributions underscore the high degree of conservation in the behavior and physiology of the central nervous system between zebrafish and humans.

In this context, the gut-brain axis is also highlighted, with a noteworthy chapter discussing the influence of microbiota on this axis and current methodologies for studying it in zebrafish.

Furthermore, this edition features a study on drug discovery using traditional Ayurvedic medicine, along with two technical chapters addressing the health and well-being of early developmental stages in housing facilities, and an essential work exploring freezing protocols for the cryopreservation of zebrafish lines.

I hope this book contributes to the continued growth of zebrafish research and inspires this generation of young researchers who seek to make a difference through their will and curiosity.

This project would not have been possible without the authors' contributions, to whom I am immensely grateful. My sincere thanks also extend to the team at IntechOpen, particularly the guest editor assistants. Finally, my editorial work on this book was supported by my employment at the Butantan Institute, through the "Conhecimento Brasil" program of the National Council for Scientific and Technological Development (CNPq).

I wish you an insightful read.

Geonildo Rodrigo Disner
Plataforma Zebrafish,
Instituto Butantan,
Sao Paulo, Brazil

Chapter 1

Introductory Chapter: A Look at the Past and Future of Zebrafish in Medical Research

Geonildo Rodrigo Disner

1. A naturalist finding

Every journey has its beginning. Therefore, the zebrafish's debut should be credited to the Scottish physician, zoologist, and botanist Francis Buchanan-Hamilton. At that time, the naturalist worked for the "*British East India Company*," which used to control extensive regions of India. Besides the commercial growth, the company sought to better understand and manage the vast territories under its control. To achieve this goal, the British company launched the endeavor called "A Statistical Survey of Bengal," which took place between 1807 and 1814. During the long expedition, Hamilton and the crew traveled through the Bengala regions (which included the parts of modern India and Bangladesh) aiming to create a complete inventory of the territory including topography, local populations, farming, and mainly natural resources.

As part of the zoology inventory and in line with his particular interest in ichthyology, Hamilton collected and documented a massive number of fish from the Ganges River and its tributaries, comprising hundreds of species unknown to occidental science. The zebrafish, first named *Cyprinus rerio,* was one of those "discoveries." Several years after the expedition, after spending a long period organizing the tremendous amount of data and specimens, it was published, in 1822, the seminal work about the found fish known as "*An Account of the Fishes Found in the River Ganges and its Branches*" [1].

In the ichthyological masterpiece, Hamilton defined the zebrafish as "*a beautiful fish I found in the Kosi river, where it grows to about two inches in length*." Moreover, he wrote: "*in an economical view, they are insipid, small, and of little value, ...nor does their number anywhere compensate for these defects*."

2. Zebrafish in the lab

About 150 years later, George Streisinger, a pioneering American molecular biologist, introduced, once and for all, the zebrafish into the biomedical research field. He recognized the zebrafish as a promising model for genetic research due to its transparent embryos, rapid development, and ease of creating and reproducing in controlled settings. At the University of Oregon, he and his colleagues established the practical methods for zebrafish, which eventually led to their widespread adoption across fields of biology, including genetics, developmental biology, and human disease research. Streisinger achieved the first ever cloning of a vertebrate species, and with his

IntechOpen

publication "*Production of clones of homozygous diploid zebra fish (Brachydanio rerio)*" he paved the way for the use of zebrafish as a genetic model organism [2].

Later, in 1988, the Monte Westerfield group first demonstrated that foreign DNA (a 5.2-kb linearized plasmid) could be integrated into the zebrafish genome and passed on *via* the germline [3, 4]. Additionally, to crown all the efforts and potential genetic and molecular applications of zebrafish, an international effort to map the fish genome began in the early 2000s, culminating in the publication of the zebrafish complete genome sequence in 2013 [5]. This was an important milestone in zebrafish history, as it solidified its global use as a vertebrate model, reinforcing its great scientific value and providing an all-around alternative to mammalian models.

3. Contributions from the past decades

Considering the pioneering research mentioned above and the deep knowledge of the zebrafish genetic composition, one of the most important contributions (in terms of forefront science) was the generation of transgenic strains through genetic manipulation.

According to The Zebrafish Information Network (ZFIN), the central database for research on *Danio rerio*, thousands of transgenic lineages have been established worldwide. This massive number of different biochemical modifications was driven primarily due to the transparent feature of the initial developmental phases of the zebrafish, which turns this animal model perfect to visual transgeny. Most transgenic lineages comprise "shiny" fish that express fluorescent proteins, such as green fluorescent protein or mCherry, tagged to specific genes or organ markers, allowing scientists to observe, in real time, detailed biological reactions within a living organism at the target site(s) of interest. Some of the most widely used applications of this biotechnology include, for example, observing nerve circuitry, immune cells attacking infections, and the action of drugs on biological substrates [6].

Another substantial contribution of the zebrafish to medical research in the past decades has been the recapitulation of human disease phenotypes. The establishment of such a variety of disease models is crucial to help our understanding of physiopathology and disease progression, not to mention the valuable contribution of these experimental platforms to test novel drugs and therapies. Indeed, the most mentioned contributions of those systems are the (1) high-throughput screening for drug discovery, (2) real-time mechanistic elucidation of the pathogenesis, besides the (3) functional validation of human mutations implicated in disease onset and progression, achieved also with the aid of the "xenographs" or "avatars" models.

The most significant importance of zebrafish models that, at least partially, reproduce human diseases lies in their ability to fill a key gap between *in vitro* studies and mammalian models, the so-called "*translational gap*." Examples of human diseases reproduced in zebrafish are hematologic diseases, cancer, infectious diseases, neurodevelopmental impairments, muscle-bone diseases, and inflammation, among others.

Moreover, in the context of the ethical principles of Reduction, Refinement, and Replacement (3Rs) in research, the zebrafish (notably at the embryo-larval stage) have been extensively used in non-clinical assays. This application is intended to test both the safety of compounds under development and efficacy performance [7].

Despite decades of dominance due to powerful genetic and diagnostic tools, inherent physiological differences between mice and humans (e.g., diet, heart rate, microbiome) call for the use of alternative animal models in parallel. In that context, zebrafish is an extraordinary experimental model for understanding the unpredictable preclinical effects of drugs under development on complex biological systems. The Thalidomide case illustrates the superior predictive power of zebrafish models. Although the drug caused severe congenital human disabilities and failed to induce defects in mice, it successfully triggered comparable morphological limb defects in zebrafish [8].

Besides the fact that chemical testing in zebrafish is faster and even easier (considering some parameters) than applying other animal models, given their metabolic similarity to mammals, zebrafish are receiving growing attention for drug screening and pharmacological studies focusing on drug absorption, distribution, metabolism, and excretion (ADME), which is due mainly to the conservation and functionality of the complex of cytochrome P450 family enzymes [9].

Last but not least, zebrafish has made substantial contributions to xenobiotic toxicity screening, notably in the context of ecotoxicology and biomonitoring, where it enables a comprehensive evaluation of the potential deleterious effects of several toxic compounds. This work has been facilitated by the abundance of standardized OECD test guidelines that recommend the use of zebrafish for testing (e.g., TG 236, 203, 210, 212, 215, 229, 230). The application of OECD-accredited tests ensures internationally comparable, reproducible, and reliable results across member countries.

4. Future horizons

As zebrafish flourishes as a valuable experimental model bridging *in vitro* and *in vivo* experimentation, the repertoire of novel approaches and laboratory techniques is continually expanding and being updated. Altogether, the zebrafish research perspectives tend toward high-resolution analysis, high-throughput screening, and direct clinical translation, solidifying its role as the premier *in vivo* system for specific biomedical questions.

The main future directions might include the translational and personalized medicine accessed by, for example, injecting patient-derived tumor cells into zebrafish embryos for oncological treatment assessment, applying CRISPR/Cas9 technology to study rare genetic diseases and create new mutations, and increasing the use of zebrafish to predict drug safety better, moving beyond just simple toxicity to complex multi-organ responses that are more relevant to humans. Other research areas that will benefit from the gathered knowledge are regenerative medicine and neuroscience. Notably, the last will be applied to better understand complex neuropsychiatric disorders that alter neural circuit function, connecting neuronal responses to other systems, such as the gut-brain link, guided by system-wide approaches.

For the next few years, translational medicine will continue to be a challenge. However, zebrafish can be part of the solution by providing a toolbox for identifying and validating therapeutic targets and candidate molecules in a vertebrate organism.

For now, the question “Is there an animal-free future in medical science?” may lack a definitive answer. Probably, we will rely on animals for a while, but in a more careful and conscious manner, firmly governed by the ethical principles of the 3Rs. The important message to take home is to maximize the use of the tremendous amount of existing animal data and to ensure that all new experimental designs adhere to a rigorously well-fitted rationale that minimizes misuse and potential suffering. Ultimately, although animal experimentation is neither desired by scientists nor society, it remains a crucial and often irreplaceable component of the scientific process dedicated to solving global health challenges and promoting human well-being.

Author details

Geonildo Rodrigo Disner[1,2]

1 Plataforma Zebrafish of the Laboratory of Applied Toxinology, Butantan Institute, São Paulo, Brazil

2 Pesquisador Conhecimento Brasil do CNPq, Brazil

*Address all correspondence to: disner.rodrigo@gmail.com

References

[1] Hamilton F. An Account of the Fishes Found in the River Ganges and its Branches. London: Constable & Co.; 1822

[2] Streisinger G, Walker C, Dower N, Knauber D, Singer F. Production of clones of homozygous diploid zebra fish (*Brachydanio rerio*). Nature. 1981;**291**(5813):293-296. DOI: 10.1038/291293a0

[3] Stuart GW, McMurray JV, Westerfield M. Replication, integration and stable germ-line transmission of foreign sequences injected into early zebrafish embryos. Development. 1988;**103**(2):403-412. DOI: 10.1242/dev.103.2.403

[4] Choe CP, Choi SY, Kee Y, Kim MJ, Kim SH, Lee Y, et al. Transgenic fluorescent zebrafish lines that have revolutionized biomedical research. Laboratory Animal Research. 2021;**37**(1):26. DOI: 10.1186/s42826-021-00103-2

[5] Howe K, Clark MD, Torroja CF, et al. The zebrafish reference genome sequence and its relationship to the human genome. Nature. 2014;**505**(7482):248

[6] Johnston HJ, Gillies SLJ, Verdon R, Stone V, Henry T, Tran L, et al. Application of transgenic zebrafish for investigating inflammatory responses to nanomaterials: Recommendations for new users. F1000Res. 2025;**12**:51. DOI: 10.12688/f1000research.128851.2

[7] Choi TY, Choi TI, Lee YR, Choe SK, Kim CH. Zebrafish as an animal model for biomedical research. Experimental & Molecular Medicine. 2021;**53**(3):310-317. DOI: 10.1038/s12276-021-00571-5. Epub 2021 Mar 1

[8] Patton EE, Zon LI, Langenau DM. Zebrafish disease models in drug discovery: From preclinical modelling to clinical trials. Nature Reviews. Drug Discovery. 2021;**20**(8):611-628. DOI: 10.1038/s41573-021-00210-8. Epub 2021 Jun 11

[9] Goldstone JV, McArthur AG, Kubota A, Zanette J, Parente T, Jönsson ME, et al. Identification and developmental expression of the full complement of cytochrome P450 genes in zebrafish. BMC Genomics. 2010;**11**:643. DOI: 10.1186/1471-2164-11-643

Chapter 2

Application of the Zebrafish Model in Neurobiological Research

Meijun Pang, Hui Zhi, Kuo Zhang, Huijie Yu, Xiaoli Gao and Xiuyun Liu

Abstract

The zebrafish (*Danio rerio*) has become an indispensable model organism in contemporary neurobiological research due to its vertebrate-specific central nervous system, high genetic conservation with humans, and suitability for high-throughput *in vivo* analysis. This chapter provides a comprehensive evaluation of zebrafish applications across key domains of neuroscience, including neural development, modeling of neurodegenerative and neuropsychiatric disorders, central nervous system regeneration, neuropharmacological drug discovery, and calcium imaging. We particularly emphasize the integration of advanced tools such as clustered regularly interspaced short palindromic repeats (CRISPR)/Cas9) gene editing, real-time calcium imaging, optogenetics, and single-cell transcriptomics to dissect neural circuitry and behavior at cellular resolution. Zebrafish models have successfully replicated core features of diseases such as Alzheimer's, Parkinson's, and autism spectrum disorder. Furthermore, their unique regenerative capacity in brain and spinal tissues offers a promising platform for uncovering the mechanisms underlying neural plasticity and repair. Despite certain translational limitations, such as the lack of a neocortex and functional divergence of gene paralogs, the zebrafish model provides a biologically relevant and ethically favorable alternative to mammalian systems. As neuroscience shifts toward systems-level, mechanism-driven inquiry, zebrafish continue to expand their impact as a versatile platform for both fundamental discovery and translational innovation in brain research.

Keywords: zebrafish model, neurobiology, neural development, CNS regeneration, neurodegenerative diseases, neuropharmacology, high-throughput screening, CRISPR, calcium imaging, translational neuroscience

1. Introduction

For decades, rodent models such as mice and rats have served as the cornerstone of neuroscience research, owing to their genetic similarity to humans and the complexity of their central nervous systems. However, intrinsic limitations—such as high maintenance costs, relatively low fecundity, lengthy reproductive cycles, and limited suitability for high-resolution *in vivo* imaging—have impeded the scalability and systemic integration of experimental neuroscience. The zebrafish has ushered in a new paradigm, offering a highly efficient, genetically tractable, and cost-effective vertebrate system for investigating neural development, modeling neurological diseases, and performing large-scale pharmacological screening. In recent years, advances in genome editing technologies (e.g., CRISPR/Cas9), high-resolution optical imaging, transgenic techniques, and automated behavioral tracking have significantly expanded the utility of zebrafish in neuroscience. Between 2005 and 2022, the number of neuroscience studies employing zebrafish has grown substantially, underscoring its rising prominence in both academic and translational contexts. Several key biological traits make zebrafish exceptionally suitable for neural research: their optical transparency during early development enables real-time *in vivo* imaging; external fertilization and high fecundity facilitate large-scale genetic manipulations; and their neuroanatomical architecture and neurotransmitter systems are evolutionarily conserved with mammals, enabling meaningful cross-species comparisons. Moreover, zebrafish exhibit a wide range of quantifiable behaviors—including anxiety, memory, aggression, and social interaction—making them a robust platform for neurobehavioral research.

This chapter reviews the multifaceted applications of zebrafish in neurobiological research, spanning from anatomical and physiological characterization of the brain to circuit-level manipulation, behavioral assays, disease modeling, and drug discovery. We explore the structural organization of the zebrafish nervous system, cutting-edge technologies for visualizing and modulating neural circuits, and standardized behavioral paradigms used to study cognitive and affective functions. Additionally, we review zebrafish models of major neurological disorders—including Alzheimer's disease, Parkinson's disease, and autism spectrum disorders—and critically evaluate their strengths and limitations in translational neuroscience. By the end of this chapter, readers will gain a deep understanding of how zebrafish have transformed modern neuroscience and why they remain a premier model for integrated brain research in the twenty-first century.

2. Advantages of zebrafish in neurobiological studies

Before delving into the specific anatomical, neurochemical, and technological aspects that underscore the utility of zebrafish in neuroscience, it is instructive to first appreciate the integrative landscape that this model organism offers. **Figure 1** illustrates the core experimental strategies and enabling technologies that make zebrafish particularly valuable for studying the nervous system. These include applications in drug screening, genetic manipulation, disease modeling, neuropharmacology, and the investigation of neurodevelopmental processes such as neuron generation and migration. Complementing these strategies are cutting-edge tools like optogenetics, light-sheet microscopy, spatial transcriptomics, and high-throughput behavioral tracking powered by artificial intelligence. Together, these features position zebrafish as a model of increasing prominence in both basic and translational neuroscience. With this integrative context in mind, the following sections will systematically explore the neuroanatomical structures, neurotransmitter systems, imaging advantages, genetic tools, and behavioral paradigms that collectively define the strengths of the zebrafish in modern neuroscience.

This figure illustrates how the zebrafish model enables an integrated experimental framework for investigating the structure and function of the vertebrate nervous system. At the core of this strategy lie the model's unique biological advantages—optical transparency, rapid ex utero development, and conserved neuroanatomy—which

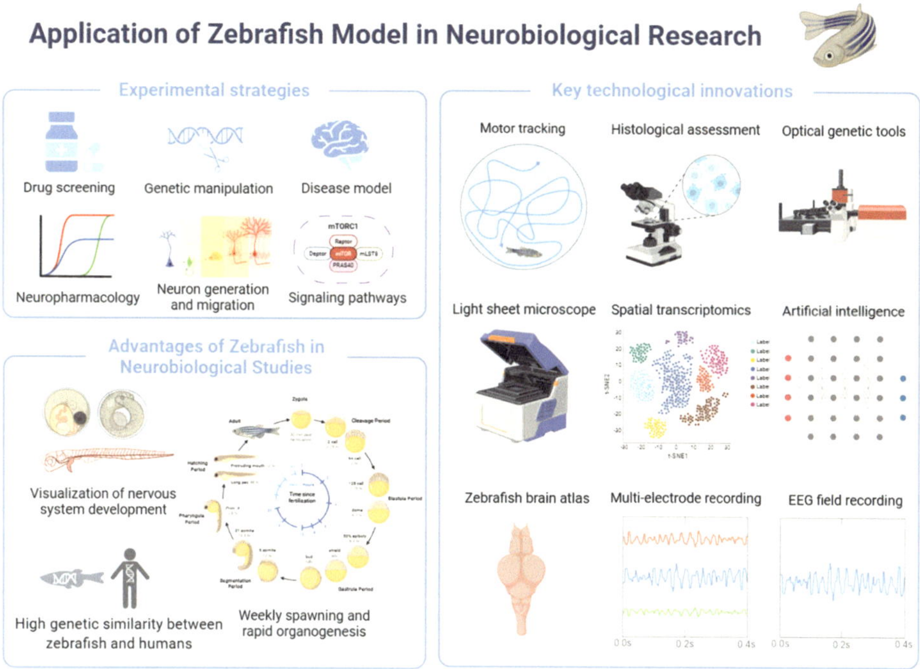

Figure 1.
Integrated experimental paradigm using the zebrafish model in neurobiological research.

collectively allow researchers to visualize and manipulate the nervous system *in vivo* in both developmental and disease contexts. These attributes have made zebrafish a powerful platform for dissecting neural circuitry, modeling neurological and psychiatric disorders, and exploring regenerative processes. Within this framework, a range of experimental strategies converge: drug discovery workflows exploit zebrafish larvae for phenotype-based screening; genetic editing tools such as CRISPR and morpholino knockdowns enable precise manipulation of neural genes; disease models recapitulate features of disorders including epilepsy, Parkinson's disease, and autism spectrum disorder; and studies of neuropharmacology assess circuit-level responses to candidate therapeutics. Supporting these strategies are advanced technologies that support high-resolution, multi-scale analysis—light-sheet microscopy for whole-brain activity imaging, optogenetics for causal circuit interrogation, spatial transcriptomics for molecular mapping, and multielectrode or electroencephalogram (EEG) field recordings for functional readouts. Furthermore, advanced behavioral platforms and artificial intelligence now permit the automated tracking and decoding of complex motor and social behaviors in freely swimming larvae, expanding the interpretive power of these systems. Anchored by references such as the zebrafish brain atlas and integrated with cross-modal data, this experimental ecosystem exemplifies a scalable and translationally relevant approach to modern neuroscience.

2.1 Neuroanatomical homology to mammals

Understanding the neuroanatomical framework of the zebrafish brain is fundamental to appreciating its value as a vertebrate model in neuroscience. Unlike invertebrate models such as *Drosophila melanogaster* and *Caenorhabditis elegans*, zebrafish possess a well-organized central nervous system (CNS) that shares significant structural and functional homology with the mammalian brain [1]. The zebrafish brain is subdivided into three primary regions: the forebrain (prosencephalon), midbrain (mesencephalon), and hindbrain (rhombencephalon)—a division conserved among vertebrates. The forebrain includes the telencephalon and diencephalon. While zebrafish lack a laminated cerebral cortex, the telencephalon contains pallial and subpallial areas that are considered functional analogs of the mammalian cortex and basal ganglia [2]. For instance, the dorsolateral pallium is believed to be homologous to the mammalian hippocampus and is implicated in spatial learning and memory. The diencephalon houses the thalamus and hypothalamus, integral to sensory integration and homeostatic regulation. The midbrain is dominated by the optic tectum—the primary visual processing center in zebrafish—functionally equivalent to the mammalian superior colliculus [3]. The hindbrain comprises the cerebellum and medulla oblongata, coordinating motor control and autonomic functions.

2.2 Conserved neurotransmitter systems

At the neurochemical level, zebrafish express highly conserved aminergic and amino acid-based neurotransmitter systems, including dopamine, serotonin (5-HT), γ-aminobutyric acid (GABA), glutamate, acetylcholine, and norepinephrine [4]. Dopaminergic neurons primarily arise from the posterior tuberculum and project to the subpallium, forming a circuitry analogous to the mammalian nigrostriatal pathway [5]. Serotonergic neurons originating in the raphe nuclei project extensively throughout the CNS, modulating mood, arousal, and anxiety. The balance of excitation and inhibition is maintained by widespread distributions of GABAergic and glutamatergic neurons, mirroring mammalian neurodynamics. These conserved systems make zebrafish a powerful model for investigating the neural basis of psychiatric and neurological disorders. Neurotoxin-based models using 1-methyl-4-phenyl-1,2,3,6-tetrahydropyridine (MPTP) or 6-hydroxydopamine effectively recapitulate Parkinsonian phenotypes through selective dopaminergic neuron degeneration [6]. Pharmacological manipulation of serotonin or GABA can reproduce anxiety-like and seizure-like behaviors, respectively, allowing drug efficacy and mechanism studies.

2.3 Adult neurogenesis and brain plasticity

One unique advantage of zebrafish is their capacity for lifelong neurogenesis. Adult zebrafish continuously generate new neurons in specific brain regions, especially the telencephalon and cerebellum, due to the persistence of radial glia-like neural progenitors. This trait sets them apart from most mammals, where neurogenesis is limited to discrete regions and diminishes with age. Following injury, zebrafish exhibit robust regenerative responses. Targeted mechanical or chemical ablation of brain regions activates neural stem cells and induces regeneration *via* Notch, Wnt, and Fibroblast Growth Factor (FGF) signaling pathways. These properties render zebrafish a valuable model for studying CNS repair mechanisms and the molecular basis of brain plasticity [3].

2.4 Unique advantages of optical transparency in *in vivo* imaging of neurodevelopment

Zebrafish embryos and larvae, particularly within the first 5–7 days postfertilization, are optically transparent. This unique feature allows noninvasive, high-resolution imaging of neural development and function at both cellular and subcellular levels. Fluorescent transgenic lines such as Tg(*elavl3:H2B-GCaMP6s*) permit pan-neuronal expression of calcium indicators for real-time monitoring of neural activity [7]. Advanced techniques like spinning-disk confocal and light-sheet microscopy enable whole-brain imaging in awake and responsive animals, facilitating unprecedented correlation between neural dynamics and behavior [8]. Additional techniques, including immunohistochemistry, in situ hybridization, and brain atlasing, complement functional imaging to create a comprehensive neuroanatomical framework [9]. Spatial transcriptomics has confirmed the anatomical-functional alignment by mapping the regional expression of neurotransmitter synthesis enzymes and synaptic transporters [2].

2.5 Optogenetics and chemogenetics in dynamic manipulation of neural circuits

The integration of optogenetic tools has further enhanced the experimental versatility of zebrafish. Channelrhodopsin-2 (ChR2), halorhodopsin (NpHR), and other opsin variants are effectively expressed in zebrafish neurons, allowing temporally precise control of neural circuits with light. These tools enable causal interrogation of neural populations in behavioral contexts, including swimming, prey capture, and sleep regulation. Optogenetic interventions are frequently paired with behavior tracking and whole-brain imaging to form closed-loop platforms for dissecting circuit function and behavioral outcomes.

2.6 Genetic accessibility and high fertility

Zebrafish possess several attributes that facilitate large-scale genetic screens. Females can produce hundreds of eggs weekly, and external fertilization allows for straightforward embryonic manipulation. Forward genetic screens using mutagens such as N-ethyl-N-nitrosourea (ENU) have identified numerous genes involved in neural development and function. Reverse genetic techniques—including morpholino-based gene knockdown and CRISPR/Cas9-mediated genome editing—enable targeted disruption of genes of interest. Importantly, these techniques can be applied at early developmental stages to observe immediate phenotypic consequences in the nervous system. The availability of a fully sequenced genome and resources like the Zebrafish Information Network (ZFIN) further supports high-throughput genomic research.

2.7 Drug permeability and pharmacological screening

Zebrafish embryos and larvae exhibit high permeability to small molecules through their skin and gills, making them ideal for pharmacological research. Compounds can be administered directly into the aquatic environment, eliminating the need for invasive delivery methods. This facilitates high-throughput drug screening for neuroactive compounds. Larval zebrafish exposed to chemical libraries have been used to identify agents that influence neural development, behavior, and neurodegenerative phenotypes. Their compatibility with 96-well plates and automation platforms enables scalable, cost-effective compound screening with behavioral readouts.

2.8 Behavioral complexity and neuropsychiatric modeling

Despite their small size and simplicity relative to mammals, zebrafish display a rich repertoire of quantifiable behaviors, including learning, memory, anxiety-like responses, social preference, and circadian activity. Studies by Collier et al. have emphasized the linkage between zebrafish behavior and underlying dopaminergic and serotonergic systems, underscoring their relevance in neuropsychiatric research [10]. Zebrafish models are extensively used to replicate symptoms of neurological and psychiatric disorders such as anxiety, autism spectrum disorder, and epilepsy. Their small size and high fecundity enable large-scale behavioral studies using automated tracking systems such as Noldus EthoVision and ViewPoint ZebraLab. These platforms support rigorous phenotyping and evaluation of genetic and pharmacological interventions. FDA-approved drugs identified or repurposed through zebrafish assays highlight the model's translational potential and relevance to human health.

The structural organization, conserved neurotransmitter systems, optical accessibility, genetic manipulability, and rich behavioral phenotypes of zebrafish collectively underscore their value in neuroscience. As the field increasingly integrates multi-scale data—from molecules to circuits to behavior—the zebrafish emerges as a versatile vertebrate system bridging the gap between invertebrate simplicity and mammalian complexity. Its suitability for both basic and translational research continues to drive innovation across developmental, behavioral, and disease-oriented neuroscience.

3. Zebrafish in neural development research

The emergence of zebrafish as a central model in developmental neurobiology is neither incidental nor merely a consequence of convenience. Unlike mammalian embryos that develop in utero, zebrafish embryos develop externally, remain optically transparent during early stages, and display high developmental synchrony. These attributes make zebrafish one of the few vertebrate organisms in which the fate of individual neural progenitor cells can be tracked *in vivo* over time with high spatial and temporal resolution. The zebrafish model encompasses all major stages of vertebrate neural development—including neural induction, neural tube formation, neural patterning, neuronal proliferation and differentiation, migration, axon guidance, synaptogenesis, and the functional integration of neural circuits—providing a comprehensive platform for the mechanistic investigation of neurogenesis [11]. Moreover, the evolutionary conservation of signaling pathways and gene regulatory networks involved in zebrafish neurodevelopment ensures the translational relevance of findings derived from this model. In particular, insights gained from zebrafish studies have contributed substantially to our understanding of human neurodevelopmental disorders, ranging from structural brain malformations to psychiatric conditions such as autism spectrum disorder (ASD).

3.1 Early neural induction and patterning: Signaling gradient and gene control

The development of the zebrafish nervous system begins with the induction of the neural plate from the ectodermal layer around 10 hours postfertilization (hpf). This structure subsequently folds to form the neural tube within 24 hours. This process is orchestrated by a finely tuned interplay of morphogen gradients and transcriptional regulation, involving key signaling pathways such as Bone Morphogenetic Protein (BMP), Wingless/Int (Wnt), Fibroblast Growth Factor (FGF), and Sonic Hedgehog (Shh).

Wnt signaling acts along the dorsoventral and anteroposterior axes, facilitating brain regionalization and the formation of the midbrain-hindbrain boundary (MHB).

Transcription factors such as *boz* and *chordin* play pivotal roles in neural induction and anterior structure specification [12]. Shh signaling, emanating from midline structures like the floor plate and notochord, is essential for ventral patterning and motor neuron differentiation. Mutations in smoothened or chemical inhibition using cyclopamine result in loss of midline structures and brain symmetry, underscoring Shh's critical role in CNS architecture.

Notch signaling regulates the balance between neural progenitor maintenance and differentiation. High Notch activity keeps progenitors undifferentiated, while reduced Notch promotes neuronal fate determination. The *mind bomb* zebrafish mutant, which lacks functional Notch signaling, exhibits excessive neuronal differentiation due to defective lateral inhibition, exemplifying the spatial coordination required during neurogenesis [13].

3.2 Regionalization of the neural tube and brain compartmentalization

By the end of the first day of development, the zebrafish neural tube is segmented into the prosencephalon (forebrain), mesencephalon (midbrain), and rhombencephalon (hindbrain), which further differentiate into specialized functional territories, including the telencephalon, diencephalon, optic tectum, cerebellum, and medulla oblongata.

Although zebrafish lack a laminated neocortex, the dorsolateral telencephalon (Dl) is considered homologous to the mammalian hippocampus. This region expresses conserved transcription factors such as *emx1* and *otx2* and shows high activity during spatial learning tasks, suggesting functional equivalence [14]. The optic tectum, which dominates the zebrafish midbrain, serves as the principal visual processing center and corresponds functionally to the mammalian superior colliculus.

3.3 Neural progenitor dynamics, neuron migration, and synaptic integration

Zebrafish neural progenitors exhibit structured proliferative behavior along the ventricular zone. These neural stem cells divide symmetrically to expand the progenitor pool or asymmetrically to produce neurons and glia. Real-time lineage tracing using *in vivo* time-lapse imaging has elucidated the detailed lineage trajectories of single progenitor cells [15]. Transgenic lines like Tg(neurog1:GFP) mark neuronal precursors, while elavl3:H2B-GCaMP6s enable calcium imaging of developing circuits at cellular resolution [16].

Within the dorsal telencephalon, progenitor zones persist into adulthood, demonstrating robust neurogenic capacity. Neural stem cells in this region respond to environmental stimuli such as learning and stress by reentering the cell cycle, a phenomenon analogous to adult hippocampal neurogenesis in mammals [17, 18]. Notch, BMP, and FGF pathways remain active in these adult niches, supporting both maintenance and regenerative responses.

3.4 Cellular mechanisms of neuronal migration and cortical organization

Neuronal migration in zebrafish mirrors mechanisms observed in mammals. Radial glial cells serve as scaffolds for radial migration, particularly in layered structures. Mutations in genes such as reelin or dcx disrupt proper migration and cortical organization, enabling the modeling of lissencephaly and other human cortical malformations in zebrafish [19, 20]. Zebrafish mutants such as valentino have contributed to understanding the genetic basis of midbrain-hindbrain boundary formation. Loss of *fgf8* in these mutants leads to disrupted neuroepithelial organization and impaired neuronal positioning, confirming its role as a critical morphogen for brain regional identity [21].

Taken together, the zebrafish provides an experimentally accessible, genetically tractable, and evolutionarily relevant model for studying neural development. The conservation of developmental signaling pathways, coupled with the ability to visualize dynamic processes *in vivo*, makes zebrafish uniquely suited to dissecting the spatiotemporal regulation of neurogenesis. Moreover, the zebrafish has proven invaluable for modeling human neurodevelopmental disorders. Gene editing technologies like CRISPR/Cas9 enable the generation of disease-relevant mutants, while advanced imaging and transcriptomic profiling provide insight into the cellular and molecular pathogenesis of these conditions. As developmental neuroscience moves toward integrative approaches encompassing cell lineage, circuit function, and behavioral outputs, the zebrafish model remains at the forefront of discovery.

4. Zebrafish in neurodegenerative disease modeling and neuropharmacological discovery: An integrated perspective

Neurodegenerative and neurodevelopmental disorders, including Alzheimer's disease (AD), Parkinson's disease (PD), epilepsy, and ASD, pose critical global public health challenges. These conditions are characterized by intricate pathological mechanisms, such as synaptic dysfunction, protein misfolding, mitochondrial dyshomeostasis, neuroinflammation, and extensive neuronal network disruption. As shown in **Figure 2**, while traditional models have been instrumental in modeling certain pathological processes, they are constrained by long developmental timelines, high maintenance costs, and limited scalability for high-throughput screening. These limitations have driven the emergence of zebrafish as a new-generation vertebrate platform for neurodegenerative research and CNS drug screening, owing to its advantages of high genetic homology, optical transparency, high fecundity, and complex, quantifiable behaviors [22].

Traditional drug development has long relied on cell cultures and rodent models, which, while useful for target validation and efficacy testing, suffer from long experimental cycles, high costs, low throughput, and limited translational relevance—particularly in the context of CNS disorders, where rapid behavioral and systems-level evaluation is crucial. In contrast, zebrafish offer a powerful alternative due to their optical transparency, rapid embryonic development, high fecundity, and quantifiable behavioral outputs. Leveraging multiwell plate formats, automated tracking, and image analysis, researchers can perform large-scale *in vivo* compound screening across thousands of larvae, assessing both efficacy and toxicity in parallel. Furthermore, zebrafish models support mechanistic validation through transgenic lines and high-resolution imaging, significantly enhancing the efficiency and translational potential of neuropharmacological discovery.

4.1 Alzheimer's disease models

Zebrafish models of AD primarily target amyloid-beta (Aβ) aggregation and tau hyperphosphorylation, which are defining histopathological features of the disease. Transgenic lines expressing mutant human amyloid precursor protein (APP) (e.g.,

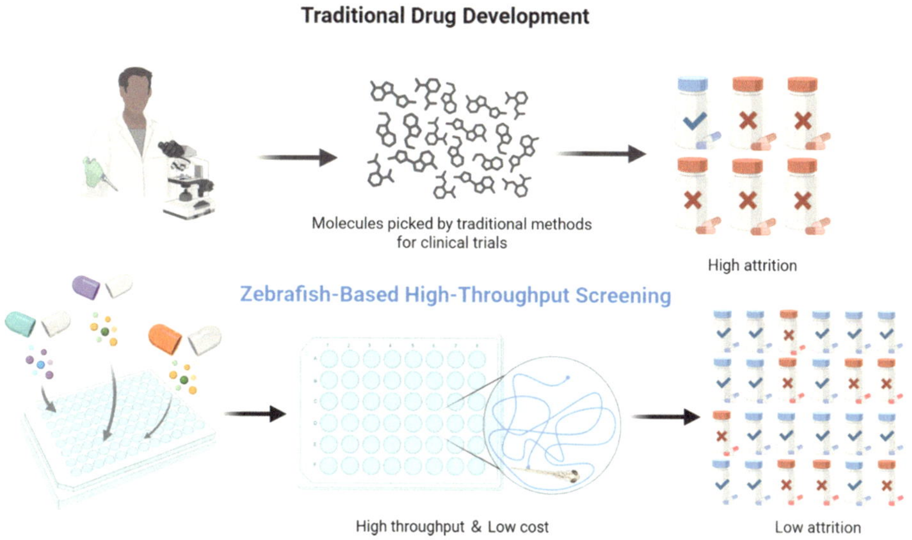

Figure 2.
Comparison of traditional drug development processes with zebrafish high-throughput drug screening platforms.

APPswe), Aβ42 peptides, and microtubule-associated protein tau (MAPT) variants (e.g., P301L tau) have been used to replicate plaque deposition and neuronal degeneration [23]. These models exhibit phenotypes consistent with human AD, including cognitive impairment, disrupted circadian rhythms, and synaptic dysfunction. Notably, Cosacak et al. performed single-cell RNA-seq (scRNA-seq) on Aβ-overexpressing zebrafish and revealed altered neurogenic lineage decisions, with neurogenic progenitors exhibiting increased heterogeneity and reduced neuronal output [24]. The impaired neurogenesis observed paralleled the reduction in hippocampal neurogenesis seen in human AD.

Behaviorally, AD-model zebrafish show impaired spatial learning (e.g., T-maze and inhibitory avoidance tasks) and heightened anxiety-like behavior (e.g., increased bottom dwelling in novel tank diving tests), consistent with early noncognitive symptoms in AD patients. Furthermore, Mhalhel et al. demonstrated that treatment with flavonoids reversed transcriptomic changes associated with synaptic proteins and energy metabolism, supporting the use of zebrafish in mechanism-based drug screening [25]. The hAPP-GFP transgene has also allowed real-time visualization of amyloid plaques *in vivo*, enabling high-resolution studies on Aβ deposition and its correlation with circuit disruption [26].

4.2 Parkinson's disease models

Zebrafish models of PD encompass both neurotoxic and genetic approaches. Neurotoxins such as MPTP, 6-OHDA, and rotenone are employed to induce dopaminergic neuronal loss *via* mitochondrial complex I inhibition and oxidative stress [27, 28]. These treatments produce hallmark PD phenotypes, including reduced spontaneous swimming, postural instability, and decreased TH+ neuron counts in the posterior tuberculum. Genetic models expressing human PD mutations in LRRK2 (G2019S), PINK1, DJ-1, and synuclein alpha (SNCA) also reproduce dopaminergic circuit impairment and behavioral motor phenotypes. Importantly, *lrrk2*-G2019S zebrafish lines reveal defects in autophagic flux and proteostasis, offering a tractable system for testing autophagy-modulating therapeutics [23, 29, 30].

Therapeutically, losartan and olmesartan—angiotensin receptor blockers—have demonstrated protective effects in MPTP-induced zebrafish PD models by stabilizing mitochondrial membrane potential and normalizing locomotion metrics [31]. These findings suggest the potential for drug repurposing in PD and support zebrafish as a mechanistic preclinical platform.

4.3 Epilepsy and excitotoxicity models

Epilepsy modeling in zebrafish commonly utilizes chemical convulsants like pentylenetetrazole (PTZ), which antagonizes GABA-A receptors, causing neural hyperexcitability. PTZ-treated larvae exhibit seizure-like behaviors such as burst swimming, tremors, and corkscrew-like rotations, which can be quantitatively analyzed *via* platforms such as EthoVision XT [32]. Additional models using kainic acid or glutamate analogs induce excitotoxicity and neurodegeneration [33]. Genetic models with mutations in ion channel genes like *scn1a* and *kcnq2* mimic specific human epilepsy syndromes. For instance, retigabine (a potassium voltage-gated channel subfamily Q (KCNQ) opener) significantly improves electrophysiological and behavioral outcomes in *kcnq2*-deficient zebrafish larvae, suggesting therapeutic relevance for early-onset epilepsy [34]. The translational value of zebrafish in drug discovery has been well demonstrated. The PTZ-induced seizure model, developed by the Peterson lab, identifies antiepileptic compounds through quantifiable hyperactivity and convulsive behaviors—yielding hits like levetiracetam and clemizole, both of which are clinically relevant [35].

4.4 Autism spectrum disorder models

Zebrafish models of ASD are grounded in mutations of key synaptic scaffolding and communication genes, such as *shank3*, *cntnap2*, and *mecp2*. These models recapitulate core ASD behaviors, including reduced social interaction, increased stereotypy, and altered sensory processing [36, 37]. For instance, *shank3b*-deficient zebrafish show significantly reduced conspecific preference in social shoaling assays and altered thigmotaxis behavior, which correlates with impaired forebrain and midbrain activity as measured *via* GCaMP6s-based calcium imaging [38]. These behavioral and functional phenotypes mirror human ASD endophenotypes and provide a foundation for high-throughput screening of social behavior-modulating drugs. Advanced imaging systems have made it feasible to track the activity of hundreds of neurons across the zebrafish brain during social exposure. Ongoing studies also employ spatial transcriptomics to link gene expression gradients with social circuit engagement [39]. Pharmacologically, treatment with oxytocin analogs, aripiprazole, and risperidone has shown partial rescue of ASD-related behaviors in zebrafish, validating predictive pharmacodynamic modeling [40, 41].

5. Zebrafish in neural regeneration and plasticity studies

Unlike mammals, whose CNS exhibits limited regenerative capacity, zebrafish possess the remarkable ability to fully repair damaged CNS structures, including the retina, spinal cord, and brain. This unique capacity has made zebrafish a pivotal model for uncovering mechanisms of neuroregeneration and neural plasticity. As neurodegenerative and traumatic CNS disorders remain largely irreversible in humans, understanding how zebrafish regenerate damaged neural circuits offers potentially transformative insights for therapeutic development.

5.1 Retinal regeneration: Müller glia as neurogenic progenitors

Zebrafish retinal regeneration is one of the earliest and most thoroughly studied examples of neural regeneration in vertebrates. Upon photoreceptor damage, Müller glial cells exit their terminally differentiated state and reenter the cell cycle, giving rise to retinal progenitor cells that subsequently differentiate into lost neuronal subtypes to reconstruct functional circuitry [42].

This regenerative response is orchestrated by key signaling pathways. Notably, Jak/Stat3 activation and *ascl1a* expression are required for the neurogenic reprogramming of Müller glia [43]. Inhibition of Notch signaling and reactivation of *Lin28* further amplify the regenerative potential, in stark contrast to the regenerative failure observed in mammalian systems [42]. These findings have inspired reprogramming strategies in murine and human Müller cells, using forced expression of Ascl1 to mimic zebrafish-like regenerative behavior [44].

5.2 Spinal cord regeneration: Cellular and molecular insights

Zebrafish can achieve complete spinal cord regeneration within 2–3 weeks after a full transection injury. This process involves ependymo-radial glial cells (ERGs), which proliferate, migrate to the lesion site, and differentiate into new neurons and glial cells [32]. Regeneration is supported by upregulation of key transcription factors such as *Sox2*, Nestin, *Stat3*, and GFAP [45]. Moreover, the immune response differs from that in mammals: microglia and macrophages in zebrafish help establish a pro-regenerative microenvironment by secreting IL-4, TGF-β, and TNF-α, promoting rather than inhibiting repair [46].

5.3 Brain regeneration: Regional neurogenesis and plasticity

Distinct from adult mammals, which restrict neurogenesis to a few niches (e.g., subventricular zone, hippocampus), adult zebrafish exhibit persistent neurogenesis across multiple brain regions, including the dorsal telencephalon, diencephalon, optic tectum, and cerebellum [47]. Radial glial cells act as neural stem cells in these zones. Following injury or neurotoxin exposure, they undergo activation, proliferation, and differentiation into functional neurons. Key signaling pathways such as Notch, Wnt, FGF, and Hedgehog play tightly regulated roles in this regenerative cascade [47]. Spatial transcriptomics and single-cell lineage tracing studies have revealed that these progenitor populations are spatially heterogeneous and possess distinct lineage potentials. For example, radial glia in the anterior dorsal telencephalon display a heightened regenerative response post-injury and can integrate their progeny into preexisting neural circuits to restore function.

5.4 Functional recovery and network rewiring

At the systems level, zebrafish exhibit extraordinary plasticity following CNS injury. Electrophysiological recordings and calcium imaging have shown that newborn neurons can rapidly establish synaptic connections and restore the temporal coordination of network activity—even in partially damaged circuits [48]. Behavioral paradigms such as free-swimming assays, optokinetic reflexes, and somite-based locomotor tests quantitatively demonstrate recovery of motor and sensory functions. The speed and efficiency of neural circuit reconstruction in zebrafish suggest the existence of unique molecular mediators or wiring rules that facilitate successful reintegration—areas that remain fertile ground for further investigation.

6. *In vivo* calcium imaging in zebrafish: A window into neural dynamics

As a critical tool in modern systems neuroscience, calcium imaging technology has demonstrated unique advantages in the zebrafish model. Due to their transparent brain structure, ease of genetic manipulation, and highly plastic neural network, zebrafish provide an ideal platform for *in vivo* observation of neuronal activity. By expressing calcium indicators (such as the genetically encoded calcium indicator based on green fluorescent protein (GFP), calmodulin, and M13 peptide (GCaMP) series), researchers can record changes in neuronal calcium ion concentration at subcellular resolution to infer electrical activity. In recent years, calcium imaging in the zebrafish nervous system has expanded from immobilized to freely behaving states, not only advancing our understanding of behaviors such as perception, movement, and learning but also providing new quantitative metrics for neurological disease modeling and drug intervention.

6.1 Whole-brain imaging in immobilized larval zebrafish: Circuit-level analysis at cellular resolution

Whole-brain calcium imaging in immobilized zebrafish larvae is typically conducted by embedding the animals in low-melting agarose and anesthetizing them. Early efforts used confocal and two-photon microscopy for region-specific imaging. However, these methods were limited by their speed and depth, preventing brain-wide simultaneous recordings. Ahrens et al. first utilized light-sheet fluorescence microscopy (LSFM) to image >80% of neurons in zebrafish larvae at ~1 Hz resolution, marking a major breakthrough [16, 49]. Vladimirov et al. improved this setup using a high-speed scanned light-sheet microscope with sCMOS detectors, capturing up to 10,000 neurons at 5–10 Hz temporal resolution [50]. Combined with a pan-neuronal expression of GCaMP6s, the platform allowed for high-resolution mapping of both evoked and spontaneous activity patterns.

The GCaMP family, especially GCaMP6s, is the standard calcium indicator in zebrafish due to its high signal-to-noise ratio and photostability. The Tg(*elavl3:H2B-GCaMP6s*) transgenic line localizes calcium signals to the nucleus, minimizing motion-induced artifacts [51]. Further specificity is achieved through genetic tools such as Gal4-UAS or Cre-Lox systems. These enable cell-type or region-specific expression of GCaMP in excitatory (*vglut2a:Gal4*) or inhibitory (*gad1b:Gal4*) neurons. Enhancer trap strategies have also been used to visualize sparse neuronal subsets [52].

Kunst et al. developed the Z-Brain anatomical template, offering a standardized 3D reference brain for zebrafish imaging studies. Calcium imaging data can be non-rigidly registered to this template for cross-sample comparisons and anatomical annotation [53]. By mapping activity induced by light, temperature, tactile, or chemical stimuli onto the Z-Brain, researchers have identified functionally responsive domains. For example, coordinated firing in the preoptic area and tectum in response to heat suggests their role in thermotactic behavior.

Even without external stimuli, the zebrafish brain exhibits rich spontaneous neural activity. Naumann et al. revealed large-scale synchronous activity between the telencephalon and diencephalon, using graph-theoretical analyses to reconstruct resting-state functional connectivity networks [54]. These analyses incorporate calcium signal time series with Pearson correlation matrices and network centrality indices, revealing small-world properties and hub architecture, despite the absence of a layered neocortex.

Imaging under immobilized conditions has also been applied in behavioral paradigms using virtual reality (VR) to simulate motion cues. DeMarco et al. used a posture-coupled VR setup to show enhanced activity in the Dl (hippocampal analog) region during navigation learning [55]. In mutant models such as CNTNAP2-deficient zebrafish, calcium imaging reveals attenuated responses in thalamic and hypothalamic regions during social stimuli, reflecting impaired information processing [56].

6.2 Imaging in freely behaving zebrafish: Dynamic integration of brain and behavior

Calcium imaging in freely behaving states aims to address a core question: how to capture real-time dynamic responses of neural circuits during an animal's active perception, exploration, or escape behaviors. Such imaging typically employs high-frame-rate, low-phototoxicity imaging systems, complemented by machine vision for real-time correction of motion artifacts. To mitigate focal drift and image blur caused by free movement, researchers often use techniques such as high-speed dual-lens synchronous capture, speckle tracking calibration, and 3D behavioral localization to stabilize and trace the imaging region.

One of the most widely used approaches is the employment of virtual reality (VR) or "fictive behavior" setups, where larvae are embedded in low-melting-point agarose with only their tail exposed for electrophysiological recording or fictive movement, while the brain remains stationary for high-resolution imaging. In this paradigm, fictive swim signals (muscle or nerve activity) are used to drive visual feedback *via* closed-loop control, enabling the study of sensorimotor integration [50].

For instance, Ahrens et al. used a two-photon laser scanning microscope and GCaMP6s-based pan-neuronal expression to acquire whole-brain calcium activity in zebrafish larvae during visual stimulation. They identified task-related, distributed functional modules spanning the hindbrain, midbrain, and tectum, suggesting an intrinsic network organization for motion processing [16]. This study laid the foundation for the "fictive swim + whole-brain imaging" framework. Further technological innovations include combining light-sheet microscopy with behavioral monitoring platforms to achieve imaging during semi-free swimming. Naumann et al. developed an integrated setup that mildly restrains the fish and presents virtual reality stimuli to the visual field. With this, they reconstructed the activity patterns underlying optomotor responses and trained a forward model that predicts motor intent from neural population activity, providing a new tool for exploring motor planning and spatial navigation [57].

To reduce behavioral artifacts due to restraint, Förster et al. introduced a semi-constrained imaging platform that uses micro-objective lenses and flexible head immobilization, enabling localized behaviors such as escape reflexes while maintaining brain stability. These platforms revealed region-specific neuronal encoding of external stimuli [58]. Technologically, the development of next-generation indicators (e.g., jGCaMP8s) and fast CMOS-based sensors significantly improved imaging resolution and fidelity. In 2022, the Del Bene lab introduced a hybrid EEG-calcium imaging system that simultaneously captured field potentials and region-specific calcium activity, allowing precise temporal correlation between electrophysiological events and local neuronal firing. This system provides valuable insights for studies of epilepsy and slow-wave sleep dynamics. Recent studies also extend into group behavior and social network imaging. Researchers employed genetic labeling and optogenetic tools to image neuronal populations in the dorsal diencephalon during social preference tasks in freely swimming zebrafish. They identified specific ensembles activated during social approach behavior, indicating that this region plays a critical role in collective behavior and social cognition, supporting zebrafish as a model for ASD [59].

In summary, although challenges such as optical distortion, motion artifacts, and imaging duration persist, the integration of advanced indicators, real-time behavioral tracking, and synchronized stimulation systems is rapidly propelling calcium imaging in freely behaving zebrafish toward a comprehensive, brain-wide framework for mapping causal circuit-behavior relationships.

7. Limitations and challenges

7.1 Anatomical differences from mammals and challenges in behavioral standardization

While zebrafish serve as powerful models in neurobiology, their anatomical differences from mammals pose notable limitations. Most critically, zebrafish lack a laminar neocortex—a structure central to higher cognition, emotion regulation, and decision-making. Although regions such as the Dl zone of the telencephalon are considered homologous to the mammalian hippocampus, the hierarchical network architecture remains insufficiently conserved [60]. Thus, zebrafish are better suited for early-phase screening and conserved mechanistic studies than for modeling high-level cognitive processes.

Behavioral complexity in zebrafish is inherently limited. Although assays exist for anxiety, novelty seeking, aggression, and social preference, zebrafish cannot model advanced phenomena such as empathy, complex decision-making, or vocal communication. Consequently, models for disorders like schizophrenia or bipolar disorder can only capture rudimentary phenotypes. Many behaviors measured are considered intermediate phenotypes, not direct analogs of human symptoms [61, 62].

7.2 Genetic redundancy and imaging limitations beyond larval stages

Zebrafish underwent a whole-genome duplication event, leading to the presence of ohnologs—gene duplicates with overlapping but distinct functions. This redundancy often results in phenotypic compensation, complicating gene knockout interpretations [63]. For instance, simultaneous knockout of *ascl1a/ascl1b* is necessary to reveal neurogenesis defects, increasing experimental complexity [64, 65]. While the zebrafish model offers a "transparency advantage" for optical imaging, this property is limited to the early developmental stages (3–7 days postfertilization, dpf). In adult zebrafish, imaging becomes significantly more challenging due to pigment deposition and skeletal ossification, necessitating the use of pigment-deficient mutants such as Casper or nacre to maintain transparency. However, these strains exhibit differences from wild-type zebrafish in physiological stability, immune system function, and neurodevelopment, which may introduce systematic biases. For example, although whole-brain calcium imaging in Casper lines can yield clear images, the lack of melanin protection in these lines makes them vulnerable to environmental stress, thereby complicating the interpretation of behavioral output results. Additionally, the application of optogenetics and chemogenetics in zebrafish is constrained by the limited number and specificity of genetic driver tools. For instance, many Cre lines still lack regional or cell-type-specific expression capabilities [65].

7.3 Translational gaps and pharmacokinetics

Clinical translatability remains a significant challenge in zebrafish research. Furthermore, in the construction of disease models, zebrafish face inherent difficulties in mimicking certain human neurodegenerative diseases. For example, the expression and deposition patterns of neurofibrillary tangles (formed by phosphorylated tau protein) and β-amyloid plaques (Aβ)—typical hallmarks of Alzheimer's disease—differ significantly between zebrafish and the human brain [66]. Although multiple research groups have developed transgenic zebrafish lines expressing human Aβ42, mutant APP, or P301L tau protein, the neurodegenerative processes in these models progress more slowly compared to mouse models, manifesting as mild cognitive impairment rather than severe neurodegeneration [67]. Additionally, the robust neuroregenerative capacity of zebrafish can interfere with the natural evolution of neuropathological processes, making it difficult to accurately recapitulate the pathological progression of certain chronic degenerative diseases. Although several compounds identified through zebrafish screening have entered preclinical development (e.g., the antiepileptic drug retigabine or the antipsychotic candidate NE-100), the overall success rate remains low. This is partly attributed to the significant differences in drug pharmacokinetics between zebrafish and mammals. Aqueous exposure cannot precisely control the effective concentration of compounds in plasma or brain tissue, leading to shifts in dose-response relationships during translation. Additionally, the zebrafish blood-brain barrier (BBB) is relatively permissive and lacks the strict drug barrier control mechanisms of mammals, such that compounds effective in zebrafish may fail in mammalian experiments due to penetration barriers.

Another concern is the overinterpretation of zebrafish data in translational research. While the zebrafish model can elucidate conserved mechanisms and screen therapeutic candidates, translating efficacy from fish to humans is nontrivial. Researchers and reviewers must avoid "overextension of the model," i.e., treating zebrafish results as definitive conclusions about human biology without appropriate validation in mammalian systems.

While zebrafish remain one of the most ethically flexible and experimentally powerful vertebrate models, their limitations warrant cautious interpretation. They are ideal for early-phase mechanistic studies and compound screening but should be complemented by mammalian models for advanced neurophysiological and translational research.

8. Future perspectives

As neuroscience advances toward a more integrated understanding of molecular mechanisms, neural circuits, behavioral phenotypes, and disease etiologies, the zebrafish model is poised to play a central role across multiple research domains. From its original use in developmental studies, behavioral analysis, and pharmacological screening, zebrafish is now at the threshold of transforming neuroscience into a multi-scale, interdisciplinary, mechanism-driven enterprise.

8.1 Single-cell omics and circuit-specific reconstruction

Recent innovations in single-cell omics are revolutionizing our understanding of brain complexity. In zebrafish, techniques such as scRNA-seq, snATAC-seq, and spatial transcriptomics enable unprecedented resolution in profiling cell lineages, state transitions, and functional heterogeneity. For instance, Cosacak et al. applied single-cell transcriptomics to an Alzheimer's disease zebrafish model and revealed plasticity reprogramming within neural stem cell populations under disease conditions [68]. Coupling these datasets with calcium imaging and optogenetics opens pathways to causally reconstruct circuit-level dynamics at the cellular atlas scale.

8.2 AI-driven neurobehavioral mapping

The convergence of artificial intelligence (AI) and computational neuroscience is expanding the utility of zebrafish as a behavioral model. High-throughput behavioral tracking systems such as Zebrabox and DeepLabCut enable precise quantification of exploratory, social, and cognitive behaviors. Studies are beginning to integrate behavioral data with neural activity using deep learning frameworks, producing neuro-behavioral maps that model task-specific trajectories and their neural correlates [69]. Such data fusion strategies are pivotal for modeling complex psychiatric disorders with heterogeneous phenotypic signatures.

8.3 System-level integration and next-generation disease modeling

The next phase of zebrafish neuroscience will be fundamentally systemic and integrative. Multimodal platforms such as Z-Brain and MapZebrain combine gene expression, behavior, and circuit mapping to support reproducible, cross-lab comparative studies. These tools facilitate the development of a dynamic neurobehavioral connectome, linking static anatomy to functional circuit engagement. Connectomics, bolstered by electron microscopy reconstruction, is already producing complete larval brain synapse maps. Open-access repositories such as zebrafish brain browser (ZBB) and Z-Brain are becoming central hubs for collaborative circuit neuroscience [70, 71].

Future disease models in zebrafish will increasingly embrace the network pathology paradigm, shifting away from monogenic constructs toward multigenic, environment-interacting frameworks. Emerging studies incorporate epigenetic landscapes such as DNA methylation to elucidate links between regulatory states and behavioral phenotypes. Concepts like zebrafish brain region transplantation or gene-behavior plasticity time-window analysis may yield new insights into disease onset and intervention timing. In contrast to mammals, zebrafish possess extraordinary regenerative capacities in the CNS, including the spinal cord, retina, and telencephalon. Mechanisms such as *stat3/ascl1a* signaling and Notch deactivation are now being evaluated for therapeutic activation in mammalian models [72]. Translating zebrafish-derived regeneration strategies to mammals could redefine clinical approaches to neural injury.

8.4 Application of electroencephalography (EEG) in zebrafish neurobiological research

Electroencephalography (EEG) is increasingly being incorporated into zebrafish neurobiological studies to provide noninvasive, real-time insights into neural oscillatory activity at the whole-brain level. Although traditionally applied to mammalian models and clinical diagnostics, recent adaptations of EEG technology to zebrafish—particularly larvae and juveniles—have enabled the detection of cortical-like electrical patterns associated with seizure states, sleep-wake transitions, and pharmacological responses. These advances have been facilitated by the development of miniaturized, high-sensitivity multichannel electrodes and transparent electrode arrays compatible with the small body size and optical access of zebrafish [73, 74].

Studies employing EEG in zebrafish have demonstrated its value in validating behavioral phenotypes with electrophysiological correlates, for example, by correlating PTZ-induced seizure-like movements with increased spike frequency and epileptiform discharges [32]. Moreover, EEG recordings are being combined with pharmacological interventions to assess neural network stabilization following drug exposure, thus providing a translationally relevant biomarker for CNS drug efficacy [75, 76]. As EEG recording in zebrafish matures, future studies are likely to integrate it with calcium imaging and optogenetics to bridge neural activity at cellular, circuit, and systems levels.

Nonetheless, limitations remain. Signal resolution is inherently constrained by the size of the larval brain and the aquatic recording environment, which can introduce electrical artifacts. Additionally, while zebrafish do not possess a neocortex, their pallial structures generate rhythmic activity patterns that appear evolutionarily conserved, but the exact correspondence to mammalian EEG rhythms (such as alpha, beta, or gamma bands) remains under investigation. To enhance the reliability and reproducibility of EEG findings, further standardization of electrode placement, signal processing pipelines, and behavioral synchronization protocols is needed.

In the broader context of neurotechnology integration, zebrafish EEG platforms are expected to benefit from developments in flexible bioelectronics, machine learning-based pattern recognition, and high-density electrode fabrication. These advances could eventually enable closed-loop neurostimulation paradigms or neurofeedback experiments in zebrafish, mirroring approaches now being explored in rodent models. As such, EEG holds promise not only as a functional readout for drug screening and disease modeling but also as a tool for probing the fundamental dynamics of vertebrate brain function in an ethically favorable and genetically tractable organism.

9. Conclusion

Over the past two decades, zebrafish have emerged from a niche developmental model to become a pillar of modern neuroscience. Their unique combination of genetic tractability, optical accessibility, behavioral quantifiability, and physiological relevance positions them as ideal vertebrate systems for multi-scale investigation. As this chapter has shown, zebrafish contribute meaningfully to our understanding of neural development, neurodegenerative diseases, neuroregeneration, and neuropharmacology. Despite inherent limitations—such as their lack of a neocortex, behavioral simplifications, and translational constraints—the zebrafish model continues to adapt through technological innovations and systemic integration. As future research converges across single-cell biology, behavioral neuroscience, and computational modeling, zebrafish will increasingly facilitate precision diagnostics, mechanism-guided therapeutics, and high-throughput behavioral analytics. In the era of data-rich, ethically mindful, and mechanism-oriented neuroscience, zebrafish are not merely experimental organisms—they are central agents of a scientific paradigm shift.

Their role is no longer auxiliary but foundational, bridging the gap between cellular detail and behavioral complexity, between model systems and human relevance.

Acknowledgements

The authors sincerely thank all colleagues and collaborators for their valuable insights and technical support throughout this work. Their constructive discussions laid a strong foundation for the preparation of this manuscript. In particular, we gratefully acknowledge the team at the Laboratory of Anesthesiology and Critical Care Medicine for their contributions to zebrafish behavioral assays and neuroimaging experiments, which provided methodological references for this study. We also extend our appreciation to the developers and maintainers of open-access zebrafish databases such as ZFIN and MapZebrain. Their high-quality public datasets enabled the integrated analysis of molecular and circuit-level findings throughout this research. Finally, we are deeply grateful to the anonymous reviewers whose constructive feedback significantly enhanced the academic rigor and clarity of this manuscript.

Conflict of interest

The authors declare no conflict of interest.

Author details

Meijun Pang[1,2*†], Hui Zhi[1†], Kuo Zhang[1,2], Huijie Yu[3], Xiaoli Gao[3] and Xiuyun Liu[1,2,4*]

1 State Key Laboratory of Advanced Medical Materials and Devices, Medical School, Tianjin University, Tianjin, China

2 Haihe Laboratory of Brain-Computer Interaction and Human-Machine Integration, Tianjin, China

3 Department of Neurosurgery, Tianjin Medical University General Hospital, Tianjin, China

4 School of Pharmaceutical Science and Technology, Tianjin University, Tianjin, China

*Address all correspondence to: xiuyun_liu@tju.edu.cn and meijun.pang@tju.edu.cn

†These authors are contributed equally.

References

[1] Lawson ND. Decoding the zebrafish genome. Nature Genetics. 2022;**54**(7):917-919. Accepted manuscript. Published online 2022/07/06. DOI: 10.1038/s41588-022-01080-5

[2] Porter BA, Mueller T. The zebrafish amygdaloid complex - Functional ground plan, molecular delineation, and everted topology. Frontiers in Neuroscience. 2020;**14**:608. Accepted manuscript. Published online 2020/08/09. DOI: 10.3389/fnins.2020.00608

[3] Haynes EM, Ulland TK, Eliceiri KW. A model of discovery: The role of imaging established and emerging non-mammalian models in neuroscience. Frontiers in Molecular Neuroscience. 2022;**15**:867010. Accepted manuscript. Published online 2022/05/03. DOI: 10.3389/fnmol.2022.867010

[4] Panula P, Sallinen V, Sundvik M, et al. Modulatory neurotransmitter systems and behavior: Towards zebrafish models of neurodegenerative diseases. Zebrafish. 2006;**3**(2):235-247. Accepted manuscript. Published online 2008/02/06. DOI: 10.1089/zeb.2006.3.235

[5] Anichtchik O, Sallinen V, Peitsaro N, Panula P. Distinct structure and activity of monoamine oxidase in the brain of zebrafish (Danio rerio). The Journal of Comparative Neurology. 2006;**498**(5):593-610. Accepted manuscript. Published online 2006/08/19. DOI: 10.1002/cne.21057

[6] Panula P, Chen YC, Priyadarshini M, et al. The comparative neuroanatomy and neurochemistry of zebrafish CNS systems of relevance to human neuropsychiatric diseases. Neurobiology of Disease. 2010;**40**(1):46-57. Accepted manuscript. Published online 2010/05/18. DOI: 10.1016/j.nbd.2010.05.010

[7] Ohnesorge N, Wilzopolski J, Steinfath M, Lewejohann L, Banneke S, Heinl C. Assessment of the effect of tricaine (MS-222)-induced anesthesia on brain-wide neuronal activity of zebrafish (Danio rerio) larvae. Frontiers in Neuroscience. 2024;**18**:1456322. Accepted manuscript. Published online 2024/10/09. DOI: 10.3389/fnins.2024.1456322

[8] Turrini L, Roschi L, de Vito G, Pavone FS, Vanzi F. Imaging approaches to investigate pathophysiological mechanisms of brain disease in zebrafish. International Journal of Molecular Sciences. 2023;**24**(12):9833. Accepted manuscript. Published online 2023/06/28. DOI: 10.3390/ijms24129833

[9] Shang CF, Wang YF, Zhao MT, et al. Real-time analysis of large-scale neuronal imaging enables closed-loop investigation of neural dynamics. Nature Neuroscience. 2024;**27**(5):1014-1018. Accepted manuscript. Published online 2024/03/12. DOI: 10.1038/s41593-024-01595-6

[10] Khan KM, Collier AD, Meshalkina DA, et al. Zebrafish models in neuropsychopharmacology and CNS drug discovery. British Journal of Pharmacology. 2017;**174**(13):1925-1944. Accepted manuscript. Published online 2017/02/22. DOI: 10.1111/bph.13754

[11] Kimmel CB, Ballard WW, Kimmel SR, Ullmann B, Schilling TF. Stages of embryonic development of the zebrafish. Developmental Dynamics. 1995;**203**(3):253-310. Accepted manuscript. Published online 1995/07/01. DOI: 10.1002/aja.1002030302

[12] Lekven AC, Thorpe CJ, Waxman JS, Moon RT. Zebrafish wnt8 encodes two wnt8 proteins on a bicistronic transcript and is required for mesoderm and neurectoderm patterning. Developmental Cell. 2001;**1**(1):103-114. Accepted manuscript. Published online 2001/11/13. DOI: 10.1016/s1534-5807(01)00007-7

[13] Itoh M, Kim CH, Palardy G, et al. Mind bomb is a ubiquitin ligase that is essential for efficient activation of notch signaling by delta. Developmental Cell. 2003;**4**(1):67-82. Accepted manuscript. Published online 2003/01/18. DOI: 10.1016/s1534-5807(02)00409-4

[14] Mueller T, Dong Z, Berberoglu MA, Guo S. The dorsal pallium in zebrafish, Danio rerio (Cyprinidae, Teleostei). Brain Research. 2011;**1381**:95-105. DOI: 10.1016/j.brainres.2010.12.089

[15] McKenna A, Findlay GM, Gagnon JA, Horwitz MS, Schier AF, Shendure J. Whole-organism lineage tracing by combinatorial and cumulative genome editing. Science. 2016;**353**(6298):aaf7907. Accepted manuscript. Published online 2016/05/28. DOI: 10.1126/science.aaf7907

[16] Ahrens MB, Orger MB, Robson DN, Li JM, Keller PJ. Whole-brain functional imaging at cellular resolution using light-sheet microscopy. Nature Methods. 2013;**10**(5):413-420. Accepted manuscript. Published online 2013/03/26. DOI: 10.1038/nmeth.2434

[17] Kizil C, Dudczig S, Kyritsis N, et al. The chemokine receptor cxcr5 regulates the regenerative neurogenesis response in the adult zebrafish brain. Neural Development. 2012;7:27. Accepted manuscript. Published online 2012/07/25. DOI: 10.1186/1749-8104-7-27

[18] Kizil C, Kaslin J, Kroehne V, Brand M. Adult neurogenesis and brain regeneration in zebrafish. Developmental Neurobiology. 2012;**72**(3):429-461. Accepted manuscript. Published online 2011/05/20. DOI: 10.1002/dneu.20918

[19] Nimura T, Itoh T, Hagio H, et al. Role of Reelin in cell positioning in the cerebellum and the cerebellum-like structure in zebrafish. Developmental Biology. 2019;**455**(2):393-408. Accepted manuscript. Published online 2019/07/20. DOI: 10.1016/j.ydbio.2019.07.010

[20] Peterson SM, Zhang J, Freeman JL. Developmental reelin expression and time point-specific alterations from lead exposure in zebrafish. Neurotoxicology and Teratology. 2013;**38**:53-60. Accepted manuscript. Published online 2013/05/15. DOI: 10.1016/j.ntt.2013.04.007

[21] Moens CB, Yan YL, Appel B, Force AG, Kimmel CB. Valentino: A zebrafish gene required for normal hindbrain segmentation. Development. 1996;**122**(12):3981-3990. Accepted manuscript. Published online 1996/12/01. DOI: 10.1242/dev.122.12.3981

[22] MacRae CA, Peterson RT. Zebrafish as tools for drug discovery. Nature Reviews. Drug Discovery. 2015;**14**(10):721-731. Accepted manuscript. Published online 2015/09/12. DOI: 10.1038/nrd4627

[23] Sarasamma S, Karim A, Orengo JP. Zebrafish models of rare neurological diseases like spinocerebellar ataxias (SCAs): Advantages and limitations. Biology (Basel). 2023;**12**(10):1322. Accepted manuscript. Published online 2023/10/27. DOI: 10.3390/biology12101322

[24] White DT, Saxena MT, Mumm JS. Let's get small (and smaller): Combining

zebrafish and nanomedicine to advance neuroregenerative therapeutics. Advanced Drug Delivery Reviews. 2019;**148**:344-359. Accepted manuscript. Published online 2019/02/16. DOI: 10.1016/j.addr.2019.01.011

[25] Mhalhel K, Sicari M, Pansera L, et al. Zebrafish: A model deciphering the impact of flavonoids on neurodegenerative disorders. Cells. 2023;**12**(2):252. Accepted manuscript. Published online 2023/01/22. DOI: 10.3390/cells12020252

[26] Ibhazehiebo K, Gavrilovici C, de la Hoz CL, et al. A novel metabolism-based phenotypic drug discovery platform in zebrafish uncovers HDACs 1 and 3 as a potential combined anti-seizure drug target. Brain. 2018;**141**(3):744-761. Accepted manuscript. Published online 2018/01/27. DOI: 10.1093/brain/awx364

[27] Pitchai A, Rajaretinam RK, Freeman JL. Zebrafish as an emerging model for bioassay-guided natural product drug discovery for neurological disorders. Medicines (Basel). 2019;**6**(2):61. Accepted manuscript. Published online 2019/06/04. DOI: 10.3390/medicines6020061

[28] Anichtchik O, Diekmann H, Fleming A, Roach A, Goldsmith P, Rubinsztein DC. Loss of PINK1 function affects development and results in neurodegeneration in zebrafish. The Journal of Neuroscience. 2008;**28**(33):8199-8207. Accepted manuscript. Published online 2008/08/15. DOI: 10.1523/jneurosci.0979-08.2008

[29] Gong X, Li S, Huang J, et al. Discovery of potent LRRK2 inhibitors by ensemble virtual screening strategy and bioactivity evaluation. European Journal of Medicinal Chemistry. 2024;**279**:116812. Accepted manuscript. Published online 2024/09/07. DOI: 10.1016/j.ejmech.2024.116812

[30] Ren C, Ding Y, Wei S, et al. G2019S variation in LRRK2: An ideal model for the study of Parkinson's disease? Frontiers in Human Neuroscience. 2019;**13**:306. Accepted manuscript. Published online 2019/09/26. DOI: 10.3389/fnhum.2019.00306

[31] Kim GJ, Mo H, Liu H, et al. A zebrafish screen reveals renin-angiotensin system inhibitors as neuroprotective via mitochondrial restoration in dopamine neurons. eLife. 2021;**10**:e69795. Accepted manuscript. Published online 2021/09/23. DOI: 10.7554/eLife.69795

[32] Baraban SC, Taylor MR, Castro PA, Baier H. Pentylenetetrazole induced changes in zebrafish behavior, neural activity and c-fos expression. Neuroscience. 2005;**131**(3):759-768. Accepted manuscript. Published online 2005/02/26. DOI: 10.1016/j.neuroscience.2004.11.031

[33] de Farias ACS, de Pieri PK, Bernardo HT, Baldin SL, Dondossola ER, Rico EP. Melatonin pretreatment protects against status epilepticus, glutamate transport, and oxidative stress induced by kainic acid in zebrafish. Molecular Neurobiology. 2022;**59**(1):266-275. Accepted manuscript. Published online 2021/10/20. DOI: 10.1007/s12035-021-02579-4

[34] Chege SW, Hortopan GA, M TD, Baraban SC. Expression and function of KCNQ channels in larval zebrafish. Developmental Neurobiology. 2012;**72**(2):186-198. Accepted manuscript. Published online 2011/06/22. DOI: 10.1002/dneu.20937

[35] Baraban SC, Dinday MT, Hortopan GA. Drug screening in Scn1a

zebrafish mutant identifies clemizole as a potential Dravet syndrome treatment. Nature Communications. 2013;**4**:2410. Accepted manuscript. Published online 2013/09/05. DOI: 10.1038/ncomms3410

[36] Liu CX, Li CY, Hu CC, et al. CRISPR/Cas9-induced shank3b mutant zebrafish display autism-like behaviors. Molecular Autism. 2018;**9**:23. Accepted manuscript. Published online 2018/04/06. DOI: 10.1186/s13229-018-0204-x

[37] Tayanloo-Beik A, Hamidpour SK, Abedi M, et al. Zebrafish modeling of autism Spectrum disorders, current status and future prospective. Front. Psychiatry. 2022;**13**:911770. Accepted manuscript. Published online 2022/08/02. DOI: 10.3389/fpsyt.2022.911770

[38] Liu CX, Peng XL, Hu CC, Li CY, Li Q, Xu X. Developmental profiling of ASD-related shank3 transcripts and their differential regulation by valproic acid in zebrafish. Development Genes and Evolution. 2016;**226**(6):389-400. Accepted manuscript. Published online 2016/08/27. DOI: 10.1007/s00427-016-0561-4

[39] Chaoul V, Dib EY, Bedran J, et al. Assessing drug administration techniques in zebrafish models of neurological disease. International Journal of Molecular Sciences. 2023;**24**(19):14898. Accepted manuscript. Published online 2023/10/14. DOI: 10.3390/ijms241914898

[40] Rubbini D, Cornet C, Terriente J, Di Donato V. CRISPR meets zebrafish: Accelerating the discovery of new therapeutic targets. SLAS Discovery. 2020;**25**(6):552-567. Accepted manuscript. Published online 2020/05/29. DOI: 10.1177/2472555220926920

[41] Spikol ED, Glasgow E. Separate roles for Med12 and Wnt signaling in regulation of oxytocin expression. Biology Open. 2018;**7**(3):bio031229. Accepted manuscript. Published online 2018/03/14. DOI: 10.1242/bio.031229

[42] Wan J, Goldman D. Retina regeneration in zebrafish. Current Opinion in Genetics and Development. 2016;**40**:41-47. Accepted manuscript. Published online 2016/06/10. DOI: 10.1016/j.gde.2016.05.009

[43] Ramachandran R, Fausett BV, Goldman D. Ascl1a regulates Müller glia dedifferentiation and retinal regeneration through a Lin-28-dependent, let-7 microRNA signalling pathway. Nature Cell Biology. 2010;**12**(11):1101-1107. Accepted manuscript. Published online 2010/10/12. DOI: 10.1038/ncb2115

[44] Jorstad NL, Wilken MS, Grimes WN, et al. Stimulation of functional neuronal regeneration from Müller glia in adult mice. Nature. 2017;**548**(7665):103-107. Accepted manuscript. Published online 2017/07/27. DOI: 10.1038/nature23283

[45] Reimer MM, Sörensen I, Kuscha V, et al. Motor neuron regeneration in adult zebrafish. The Journal of Neuroscience. 2008;**28**(34):8510-8516. Accepted manuscript. Published online 2008/08/22. DOI: 10.1523/jneurosci.1189-08.2008

[46] Tsarouchas TM, Wehner D, Cavone L, et al. Dynamic control of proinflammatory cytokines Il-1β and Tnf-α by macrophages in zebrafish spinal cord regeneration. Nature Communications. 2018;**9**(1):4670. Accepted manuscript. Published online 2018/11/09. DOI: 10.1038/s41467-018-07036-w

[47] Kroehne V, Freudenreich D, Hans S, Kaslin J, Brand M. Regeneration of the adult zebrafish brain from

neurogenic radial glia-type progenitors. Development. 2011;**138**(22):4831-4841. Accepted manuscript. Published online 2011/10/19. DOI: 10.1242/dev.072587

[48] Baraban M, Anselme I, Schneider-Maunoury S, Giudicelli F. Zebrafish embryonic neurons transport messenger RNA to axons and growth cones in vivo. The Journal of Neuroscience. 2013;**33**(40):15726-15734. Accepted manuscript. Published online 2013/10/04. DOI: 10.1523/jneurosci.1510-13.2013

[49] Bhattarai P, Thomas AK, Cosacak MI, et al. IL4/STAT6 signaling activates neural stem cell proliferation and neurogenesis upon amyloid-β42 aggregation in adult zebrafish brain. Cell Reports. 2016;**17**(4):941-948. Accepted manuscript. Published online 2016/10/21. DOI: 10.1016/j.celrep.2016.09.075

[50] Vladimirov N, Mu Y, Kawashima T, et al. Light-sheet functional imaging in fictively behaving zebrafish. Nature Methods. 2014;**11**(9):883-884. Accepted manuscript. Published online 2014/07/30. DOI: 10.1038/nmeth.3040

[51] Chen TW, Wardill TJ, Sun Y, et al. Ultrasensitive fluorescent proteins for imaging neuronal activity. Nature. 2013;**499**(7458):295-300. Accepted manuscript. Published online 2013/07/23. DOI: 10.1038/nature12354

[52] Förster D, Arnold-Ammer I, Laurell E, et al. Genetic targeting and anatomical registration of neuronal populations in the zebrafish brain with a new set of BAC transgenic tools. Scientific Reports. 2017;7(1):5230. Accepted manuscript. Published online 2017/07/14. DOI: 10.1038/s41598-017-04657-x

[53] Kunst M, Laurell E, Mokayes N, et al. A cellular-resolution atlas of the larval zebrafish brain. Neuron. 2019;**103**(1):21-38.e25. Accepted manuscript. Published online 2019/05/31. DOI: 10.1016/j.neuron.2019.04.034

[54] Naumann EA, Fitzgerald JE, Dunn TW, Rihel J, Sompolinsky H, Engert F. From whole-brain data to functional circuit models: The zebrafish optomotor response. Cell. 2016;**167**(4):947-960.e920. Accepted manuscript. Published online 2016/11/05. DOI: 10.1016/j.cell.2016.10.019

[55] DeMarco E, Xu N, Baier H, Robles E. Neuron types in the zebrafish optic tectum labeled by an id2b transgene. The Journal of Comparative Neurology. 2020;**528**(7):1173-1188. Accepted manuscript. Published online 2019/11/15. DOI: 10.1002/cne.24815

[56] Dalla Vecchia E, Mortimer N, Palladino VS, et al. Cross-species models of attention-deficit/hyperactivity disorder and autism spectrum disorder: Lessons from CNTNAP2, ADGRL3, and PARK2. Psychiatric Genetics. 2019;**29**(1):1-17. Accepted manuscript. Published online 2018/10/31. DOI: 10.1097/ypg.0000000000000211

[57] Naumann EA, Kampff AR, Prober DA, Schier AF, Engert F. Monitoring neural activity with bioluminescence during natural behavior. Nature Neuroscience. 2010;**13**(4):513-520. Accepted manuscript. Published online 2010/03/23. DOI: 10.1038/nn.2518

[58] Förster D, Dal Maschio M, Laurell E, Baier H. An optogenetic toolbox for unbiased discovery of functionally connected cells in neural circuits. Nature Communications. 2017;**8**(1):116. Accepted manuscript. Published online 2017/07/26. DOI: 10.1038/s41467-017-00160-z

[59] de Polavieja GG, Orger MB. Social behavior: A neural circuit for social behavior in zebrafish. Current Biology. 2018;**28**(15):R828-R830. DOI: 10.1016/j.cub.2018.06.065

[60] Wullimann MF, Mueller T. Identification and morphogenesis of the eminentia thalami in the zebrafish. The Journal of Comparative Neurology. 2004;**471**(1):37-48. Accepted manuscript. Published online 2004/02/26. DOI: 10.1002/cne.20011

[61] Kalueff AV, Stewart AM, Gerlai R. Zebrafish as an emerging model for studying complex brain disorders. Trends in Pharmacological Sciences. 2014;**35**(2):63-75. Accepted manuscript. Published online 2014/01/15. DOI: 10.1016/j.tips.2013.12.002

[62] Kalueff AV, Echevarria DJ, Stewart AM. Gaining translational momentum: More zebrafish models for neuroscience research. Progress in Neuro-Psychopharmacology and Biological Psychiatry. 2014;**55**:1-6. Accepted manuscript. Published online 2014/03/07. DOI: 10.1016/j.pnpbp.2014.01.022

[63] Postlethwait JH, Woods IG, Ngo-Hazelett P, et al. Zebrafish comparative genomics and the origins of vertebrate chromosomes. Genome Research. 2000;**10**(12):1890-1902. Accepted manuscript. Published online 2000/12/16. DOI: 10.1101/gr.164800

[64] Park HC, Shin J, Roberts RK, Appel B. An olig2 reporter gene marks oligodendrocyte precursors in the postembryonic spinal cord of zebrafish. Developmental Dynamics. 2007;**236**(12):3402-3407. Accepted manuscript. Published online 2007/10/31. DOI: 10.1002/dvdy.21365

[65] Park SO, Lee YJ, Seki T, et al. ALK5- and TGFBR2-independent role of ALK1 in the pathogenesis of hereditary hemorrhagic telangiectasia type 2. Blood. 2008;**111**(2):633-642. Accepted manuscript. Published online 2007/10/04. DOI: 10.1182/blood-2007-08-107359

[66] Bhattarai P, Thomas AK, Cosacak MI, et al. Modeling amyloid-β42 toxicity and neurodegeneration in adult zebrafish brain. Journal of Visualized Experiments. 2017;**128**:e56014. Accepted manuscript. Published online 2017/11/21. DOI: 10.3791/56014

[67] Saleem S, Kannan RR. Zebrafish: A promising real-time model system for nanotechnology-mediated neurospecific drug delivery. Nanoscale Research Letters. 2021;**16**(1):135. Accepted manuscript. Published online 2021/08/24. DOI: 10.1186/s11671-021-03592-1

[68] Cosacak MI, Bhattarai P, Reinhardt S, et al. Single-cell transcriptomics analyses of neural stem cell heterogeneity and contextual plasticity in a zebrafish brain model of amyloid toxicity. Cell Reports. 2019;**27**(4):1307-1318.e1303. Accepted manuscript. Published online 2019/04/25. DOI: 10.1016/j.celrep.2019.03.090

[69] Fan Y-L, Hsu F-R, Wang Y, Liao L-D. Unlocking the potential of zebrafish research with artificial intelligence: Advancements in tracking, processing, and visualization. Medical and Biological Engineering and Computing. 2023;**61**(11):2797-2814. DOI: 10.1007/s11517-023-02903-1

[70] Brooks P, Champion A, Costa M. Mapping of the zebrafish brain takes shape. Nature Methods. 2022;**19**(11):1345-1346. DOI: 10.1038/s41592-022-01637-6

[71] Vogt N. Zooming into the larval zebrafish brain. Nature Methods.

2017;**14**(7):652-652. DOI: 10.1038/nmeth.4361

[72] Goldshmit Y, Sztal TE, Jusuf PR, Hall TE, Nguyen-Chi M, Currie PD. Fgf-dependent glial cell bridges facilitate spinal cord regeneration in zebrafish. The Journal of Neuroscience. 2012;**32**(22):7477-7492. Accepted manuscript. Published online 2012/06/01. DOI: 10.1523/jneurosci.0758-12.2012

[73] Kundap UP, Kumari Y, Othman I, Shaikh MF. Zebrafish as a model for epilepsy-induced cognitive dysfunction: A pharmacological, biochemical and behavioral approach. Frontiers in Pharmacology. 2017;**8**:515. Accepted manuscript. Published online 2017/08/22. DOI: 10.3389/fphar.2017.00515

[74] Lee Y, Lee KJ, Jang JW, Lee SI, Kim S. An EEG system to detect brain signals from multiple adult zebrafish. Biosensors and Bioelectronics. 2020;**164**:112315. Accepted manuscript. Published online 2020/06/20. DOI: 10.1016/j.bios.2020.112315

[75] Lee DA, Oikonomou G, Cammidge T, et al. Neuropeptide VF neurons promote sleep via the serotonergic raphe. eLife. 2020;**9**:e54491. Accepted manuscript. Published online 2020/12/19. DOI: 10.7554/eLife.54491

[76] Lee Y, Seo HW, Lee KJ, Jang JW, Kim S. A microfluidic system for stable and continuous EEG monitoring from multiple larval zebrafish. Sensors (Basel). 2020;**20**(20):5903. Accepted manuscript. Published online 2020/10/23. DOI: 10.3390/s20205903

Chapter 3

Modeling Neuropsychiatric Diseases and Drug Responses in Zebrafish

Antonella Damiano, Laura Sánchez and Álvaro Arana

Abstract

Zebrafish has emerged as a powerful vertebrate model for neuropsychiatric drug research, offering unique advantages for high-throughput screening, behavioral phenotyping, and mechanistic evaluation of therapeutic compounds. Its genetic tractability, optical transparency, and conserved neurochemical pathways enable the modeling of complex brain disorders such as autism spectrum disorders, schizophrenia, depression, and epilepsy. Exposure to neuroactive agents like pentylenetetrazole (PTZ), valproic acid, and scopolamine elicits robust, quantifiable behavioral and molecular phenotypes, including seizure-like activity, impaired social behavior, and altered expression of neuroinflammatory and neuromodulatory markers. Furthermore, zebrafish models integrate transcriptomic, proteomic, and toxicity data, enabling a comprehensive assessment of both therapeutic efficacy and safety. Automated imaging and behavioral tracking systems, along with CRISPR-based genetic models, facilitate the discovery of genotype-specific drug responses and novel neuroactive candidates. By bridging *in vitro* and mammalian models, zebrafish contribute to personalized neuropsychopharmacology and accelerate the translation of basic neuroscience into clinically actionable strategies.

Keywords: neuropsychiatric disorders, behavior analysis, drug screening, mutant models, transcriptomics, phenotyping

1. Introduction

The zebrafish (*Danio rerio*) has emerged as a robust model for studying human neurological and neuropsychiatric disorders, thanks to its remarkable anatomical and functional similarities to the mammalian central nervous system (CNS). Its brain is organized into the spinal cord, hindbrain, midbrain, and forebrain, a conserved architecture that allows the modeling of complex conditions, including neurodevelopmental, neurodegenerative, and psychiatric disorders [1, 2]. Comparative neuroanatomy shows that key regions such as the cerebellum, hypothalamus, and olfactory system share homologous structures and functions with humans (**Figure 1**). The zebrafish cerebellum, for instance, contains Purkinje and granule cells with synaptic

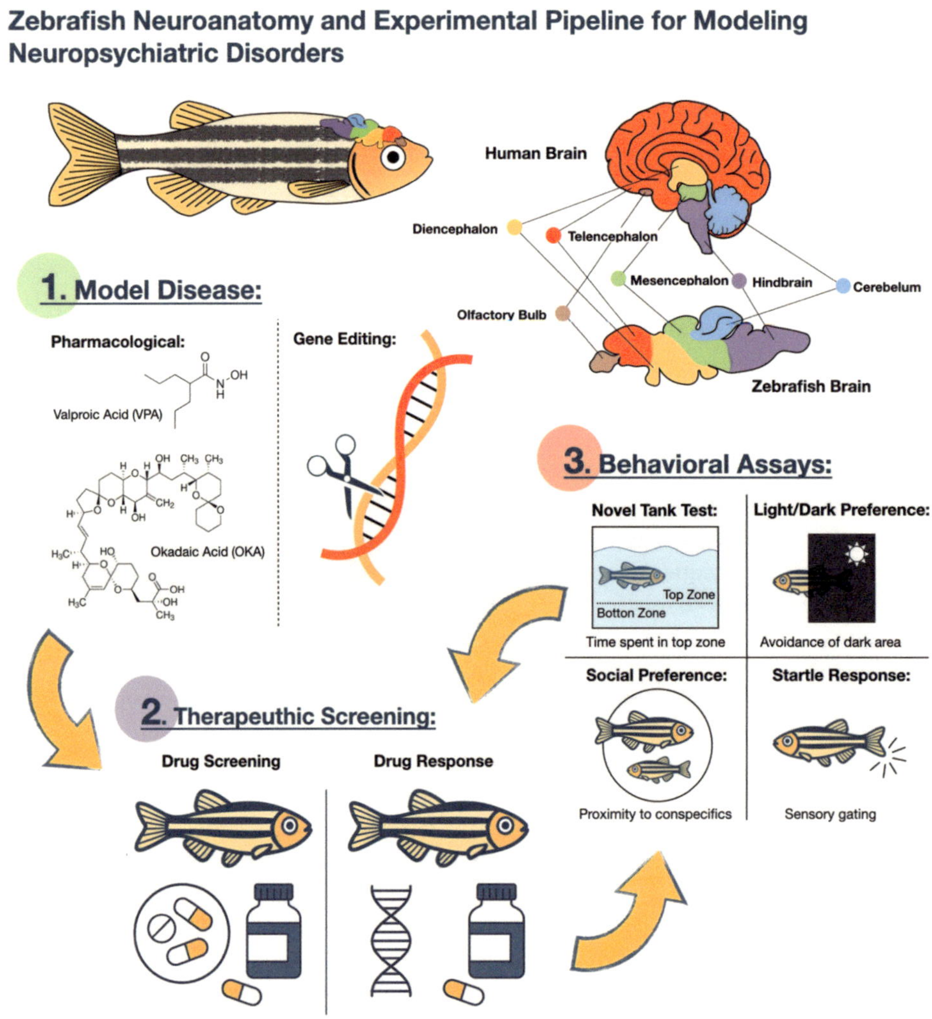

Figure 1.
Zebrafish brain anatomy and workflow for modeling and evaluating neuropsychiatric disorders. The experimental pipeline comprises three stages: (1) disease modeling via pharmacological exposure (e.g., valproic acid, okadaic acid) or genetic manipulation (e.g., CRISPR/Cas9); (2) drug screening and therapeutic evaluation; and (3) behavioral testing, including novel tank diving, light/dark preference, startle response, and social interaction assays. The anatomical conservation of key brain regions, such as the pallium, cerebellum, and diencephalon, supports the translational relevance of zebrafish for studying psychiatric conditions and testing neuroactive compounds.

connections, genetic markers, and specialized properties similar to the human cerebellar cortex [3]. One key difference lies in the positioning of projection neuron cell bodies in the cortex (rather than deep nuclei as in mammals) and in the eversion of the telencephalic hemispheres, which contrasts with the evagination seen in mammals [4, 5]. Nonetheless, neurogenesis, axonogenesis, and specific neuronal subtypes display notable similarities between zebrafish and mammals, as demonstrated by comparative neuroanatomical and neurogenetic studies [4–7].

Zebrafish are increasingly recognized in translational neuroscience due to their rapid development, genetic manipulability, optical transparency, and compatibility with live imaging and high-throughput assays [8–11]. These features have enabled their successful use in modeling neurodegenerative diseases such as Alzheimer's disease (AD), Parkinson's disease (PD), autism spectrum disorder (ASD), and amyotrophic lateral sclerosis (ALS), as well as in investigating cognition and the molecular basis of behavior [12, 13]. They are particularly valuable for identifying brain-active compounds and genes involved in learning, memory, and social function [14–16]. Their rich behavioral repertoire—including anxiety-like behavior, social interaction, and sensorimotor gating—can be robustly assessed through aquatic-based paradigms in both larvae and adults [11, 13, 16–18].

The zebrafish dopaminergic system, implicated in motor control and affected in Parkinson's disease, has been widely characterized. Dopaminergic neurons located in the olfactory bulbs, telencephalon, pretectal area, and diencephalon [19–21] regulate locomotor behavior, and their disruption through lesions or dopamine antagonists leads to motor impairment [19–27]. Neurotoxic agents such as 1-methyl-4-phenyl-1,2,3,6-tetrahydropyridine (MPTP) or 6-OHDA replicate Parkinsonian phenotypes in adults, inducing dopaminergic degeneration and reduced swimming capacity [28]. Moreover, glial cell types—including oligodendrocytes, astrocyte-like cells, and microglia—are present in zebrafish and contribute to studies on neuroinflammation, myelination, and neurodegeneration [29–33]. The existence of a functional blood-brain barrier further enhances their relevance in CNS pharmacology, allowing assessment of drug penetration and neurotoxicity [34].

Despite the prevalence of neuropsychiatric disorders, current treatments often fall short, and the underlying biological mechanisms remain elusive. Zebrafish offer a scalable and cost-effective alternative to rodent models, enabling efficient modeling of disease-relevant phenotypes through pharmacological or genetic interventions and behavioral analysis. As illustrated in **Figure 1**, their use encompasses disease induction, therapeutic screening, and behavioral testing across a wide range of CNS conditions. This chapter outlines the contribution of zebrafish models in advancing our understanding of disorders such as AD, PD, Huntington's, epilepsy, ALS, ASD, depression, anxiety, and psychosis, highlighting how genetic editing, neurochemical profiling, and behavioral phenotyping converge to support mechanistic insight and therapeutic discovery [35–52].

2. Modeling neuropsychiatric disorders in zebrafish

Zebrafish enable the study of neurodevelopmental and neuropsychiatric disorders through genetic, pharmacological, and environmental approaches. Their *ex-utero* development and optical transparency facilitate direct observation of brain formation and real-time behavioral analysis [53]. Both larval and adult zebrafish display quantifiable behaviors relevant to neuropsychiatric research, such as learning, memory, circadian rhythms, locomotion, and emotional states, which can be linked to neural mechanisms using *in vivo* imaging and optogenetics [50, 54–58]. The model also supports comprehensive molecular analyses and modeling of genetic mutations and environmental risk factors, making zebrafish a powerful platform for high-throughput screening and translational studies in neuropsychiatry [53].

2.1 Pharmacological disease models in zebrafish: Applications in CNS disorders

Pharmacological modeling in zebrafish has emerged as a valuable complement to genetic and environmental approaches for studying central nervous system (CNS) disorders. These models rely on the administration of neuroactive compounds to induce disease-relevant phenotypes by targeting neurotransmitter systems, inducing oxidative stress, or replicating key pathological features. Their high-throughput compatibility and conserved neurochemical systems make zebrafish ideal for studying pathophysiological mechanisms *in vivo*.

Numerous compounds have been validated for modeling specific CNS pathologies. For instance:

- Okadaic acid (OKA), a phosphatase inhibitor, induces tau hyperphosphorylation and amyloid-β accumulation, modeling AD.
- Scopolamine, a muscarinic acetylcholine receptor antagonist, impairs learning and memory, mimicking aspects of cholinergic dysfunction in AD [59].
- Aluminum trichloride has been used to replicate AD-like motor and cognitive impairments, including bradykinesia [60].
- Valproic acid (VPA), a histone deacetylase inhibitor, induces ASD-like phenotypes following embryonic exposure, including altered locomotion and impaired social behavior [61, 62].
- Quinpirole, a dopamine D2 receptor agonist, elicits stereotypies and behavioral rigidity relevant to obsessive-compulsive disorder (OCD) and schizophrenia [43].

Among seizure models, pentylenetetrazole (PTZ), a GABA_A receptor antagonist, is widely used to induce generalized seizure-like behavior. Larvae between 4 and 7 days postfertilization display robust, reproducible responses to PTZ, including hyperactivity and clonic-like convulsions [63].

These models reproduce key behavioral and molecular alterations of human pathologies and provide robust platforms for investigating disease mechanisms. Importantly, they establish a foundation for evaluating therapeutic efficacy, as described in Section 4.1. Together, these models underscore the versatility of zebrafish in modeling CNS disorders. They support the systematic evaluation of therapeutic candidates across behavioral, molecular, and electrophysiological dimensions and complement genetic, environmental, and multi-omic strategies discussed throughout this chapter.

2.2 Genetic manipulation and gene-environment interactions in a zebrafish model

Genetic manipulation in zebrafish has been central to modeling neurodevelopmental disorders and exploring the molecular mechanisms underlying psychiatric disease. Early studies relied heavily on morpholino antisense oligonucleotides (MOs), which transiently block mRNA translation or splicing to disrupt gene function. This approach was instrumental in the initial functional characterization of ASD-related genes. For example, knockdown of *shank3a* induced tactile hypersensitivity and microcephaly, while morpholino targeting of *auts2*, *syngap1b*, and *cep41* produced developmental and behavioral phenotypes consistent with ASD, including craniofacial abnormalities and impaired social interaction [64–70]. Despite their utility,

MOs are known to produce off-target effects such as *p53*-mediated apoptosis [8], and current recommendations advise their use only when combined with rescue assays or validated against stable knockout lines generated by genome editing. As a result, the field has progressively shifted toward more reliable and inheritable gene-editing technologies.

The introduction of CRISPR/Cas9 has revolutionized zebrafish genetics by enabling the precise and efficient generation of stable mutants. This method has been applied to model complex neuropsychiatric traits by targeting genes involved in synaptic function, ion signaling, and stress regulation. CRISPR-based disruption of cacna1c and *cacna1da*—both associated with schizophrenia—has resulted in phenotypes such as impaired prepulse inhibition (PPI), hyperactivity, and altered neurotransmission [71, 72]. Similarly, *akt3* mutants exhibit schizophrenia-relevant behavioral and molecular features, while *nr3c2*, a gene involved in stress and sleep regulation, has been linked to glucocorticoid sensitivity and anxiety-like behaviors in CRISPR-edited zebrafish [8, 73]. In an extensive CRISPR screen, Capps et al. generated 27 mutant lines targeting 17 ASD-associated genes, including *deaf1*, *nsd2*, *hectd4*, *tbl1xr1*, and *tcf7l2*, many of which showed convergent neurodevelopmental and neuropeptidergic alterations [71, 74]. Other studies have used CRISPR to model Brunner syndrome by editing *mao*, a key regulator of serotonin metabolism, which results in behavioral disinhibition and altered neurotransmitter profiles [75]. Zebrafish mutants for *stxbp1a* and *stxbp1b*, genes linked to epilepsy, display seizure susceptibility, abnormal locomotion, and comorbid cardiac dysfunction [76].

Complementary genome-editing strategies have also contributed to the expansion of zebrafish disease models. TALEN-mediated knockouts have been employed to investigate genes such as *dyrk1a*, associated with Down syndrome and ASD, revealing motor and cognitive impairments [57], and *oxtr*, a gene critical for social recognition and bonding, whose disruption leads to deficits in social preference and cohesion [64, 77]. Beyond mutational approaches, transcriptomic and in situ expression mapping have identified neuropsychiatric risk genes such as *negr1* and *opcml*, which are broadly expressed during early neurodevelopment and implicated in schizophrenia and depression [78]. These genes are thought to influence neural circuit formation and synaptic plasticity.

Together, these genetic tools have provided a versatile and scalable platform for dissecting gene function in neuropsychiatric disease. **Table 1** summarizes key examples of gene manipulations in zebrafish models, detailing the editing techniques used and their associated clinical phenotypes.

In addition to genetic tools, zebrafish provide a robust platform for studying how environmental exposures interact with genetic backgrounds to influence neurodevelopmental outcomes. A well-characterized example is ethanol exposure, which induces cerebellar hypoplasia, corpus callosum abnormalities, and persistent behavioral impairments modeling fetal alcohol spectrum disorder (FASD) [80, 81]. Strain-dependent differences in ethanol vulnerability highlight the modulatory role of genetic background in environmental neurotoxicity.

Similarly, exposure to okadaic acid (OKA) mimics Alzheimer's-like pathology through tau hyperphosphorylation and amyloid-β accumulation, demonstrating how environmental insults can trigger disease-relevant phenotypes in a genotype-sensitive manner. Additional neurotoxicants—such as aluminum trichloride, pesticides, heavy metals, and endocrine disruptors—have been used to model neurodevelopmental toxicity, often in combination with genetic manipulation or transcriptomic readouts. As summarized in **Table 2**, these experimental

Gene	Technique	Associated disorder	Ref.
shank3a	Morpholino/CRISPR	ASD	[64–66]
cacna1c	CRISPR	Schizophrenia	[71, 72]
cacna1da	CRISPR	Schizophrenia	[71, 72]
akt3	CRISPR	Schizophrenia	[73]
mao	CRISPR	Brunner syndrome	[75]
AUTS2	Morpholino	ASD	[67]
cep41	Morpholino	ASD	[68]
syngap1b	Morpholino	ASD	[79]
DYRK1A	TALEN	ASD/down syndrome	[57]
oxtr	TALEN	Social behavior deficits	[77]
nr3c2	CRISPR	Stress, sleep, ASD	[8]
stxbp1a/stxbp1b	CRISPR	Epilepsy, neurodevelopmental disorders	[76]
DEAF1 NSD2 HECTD4 TBL1XR1 TCF7L2	CRISPR	ASD	[74] [74] [74] [74] [74]

Table 1.
Key applications of zebrafish in neurodevelopmental disorder modeling.

Application domain	Techniques used	Key findings	Ref.
Gene knockdown	MOs	Transient disruption of target genes; useful for early phenotyping	[64, 65, 69]
Gene editing	CRISPR/Cas9, TALEN	Stable gene-specific mutants; essential for modeling psychiatric disorders	[57, 71, 72, 74, 77]
Expression mapping	Transcriptomics, in situ hybridization	Spatiotemporal profiling of disease-relevant gene expression	[78, 82]
Environmental toxicity	Ethanol, VPA, OKA	Models of FASD, ASD, AD; strong gene-environment interaction	[80, 81, 83]
Neurochemical modeling	Neurotransmitter targeting (e.g., MAO, GABA)	Mimics dopaminergic, serotonergic imbalance in neuropsychiatric diseases	[75, 84–86]
Behavioral phenotyping	PPI, social interaction, habituation	Endophenotypic traits for ASD, schizophrenia, anxiety	[8, 47, 87]
High-throughput screening	Immersion drug exposure + tracking systems	Efficient profiling of efficacy/toxicity of neuroactive compounds	[10, 11, 79]

Table 2.
Overview of experimental approaches and findings in zebrafish models.

paradigms span a wide array of strategies, including gene editing, neurochemical targeting, environmental modeling, and high-throughput behavioral phenotyping, offering a comprehensive toolbox for neuropsychiatric research. These approaches allow the exploration of exposome-based vulnerability, revealing how extrinsic factors like toxins or maternal health interact with intrinsic susceptibility during critical windows of brain development.

While these environmental models do not primarily serve to evaluate drug efficacy, they are instrumental in identifying mechanistic pathways linked to neuroinflammation, oxidative stress, synaptic dysregulation, and behavioral impairment—many of which are further discussed in Section 4.2. Their integration with genetic approaches strengthens the translational value of zebrafish for studying multifactorial psychiatric diseases, including ASD, schizophrenia, and FASD.

3. Therapeutic testing and drug response in zebrafish neuropsychiatric disease models

Zebrafish disease models provide a translational framework to evaluate neuropsychiatric treatments under defined pathological conditions, capturing key molecular and behavioral alterations observed in disorders such as Alzheimer's, Parkinson's, ALS, epilepsy, ASD, and stress-induced pathology. These models enable the assessment of drug effects on neuroinflammation, oxidative stress, synaptic function, and neurotrophic signaling and allow direct comparison between treated and untreated states across multiple biological levels. The use of CRISPR-generated mutants and copy number variant models facilitates pharmacogenomic studies, revealing allele-specific responses and gene-dose effects. Standardized behavioral assays, combined with molecular and imaging tools, support the identification of functional recovery and mechanistic targets. Throughout this section, we explore how zebrafish contribute to therapeutic screening, multilevel validation, and the development of precision neuropsychopharmacology, connecting phenotype, mechanism, and genotype in an integrated *in vivo* system.

3.1 Disease models and molecular hallmarks in neurodegeneration

Zebrafish have been widely used to model major neurodegenerative disorders such as AD, PD, ALS, and epilepsy. While several pharmacological models have already been described in Section 2.1, and gene-editing techniques are introduced in Section 2.2, here we focus on disease-relevant genetic models specifically applied to neurodegenerative conditions, along with their associated molecular hallmarks and utility for therapeutic screening. In AD, CRISPR-based mutations in *psen2* replicate mitochondrial dysfunction and synaptic dysregulation, mimicking key features of familial AD [88, 89]. These models complement pharmacological approaches such as OKA exposure, which promotes tau hyperphosphorylation and amyloid-β accumulation *via* GSK3β activation [83, 90]. In ALS, zebrafish expressing pathogenic variants of *sod1*, *tardbp*, *fus*, and *c9orf72* exhibit hallmark features such as axonopathy, neuromuscular junction degeneration, and impaired swimming. Notably, *sod1-g93a* and *g37r* mutants display reduced axonal branching, while *tardbp* mutants show phenotypes reversible by progranulin treatment, and *fus* mutants induce toxic gain-of-function effects [45].

In PD models, loss of dopaminergic neurons and bradykinesia have been recapitulated in zebrafish using neurotoxins such as MPTP and paraquat, the latter of which impairs dopaminergic differentiation during early development, adding insight into environmental contributions to PD risk [28]. Epilepsy is commonly modeled through PTZ, a *gabaa* receptor antagonist that induces robust seizure-like activity, detectable with genetically encoded calcium indicators like GCaMP (Genetically encoded calcium indicator), which support high-content drug screening workflows [65, 91].

Zebrafish models also allow the investigation of conserved molecular hallmarks associated with neurodegeneration, such as neuroinflammation and oxidative stress. Chronic exposure to stressors or toxicants upregulates pro-inflammatory cytokines (il-1β, tnf-α, ifnγ, il-6) and microglial markers (csf1r, cd68) [92–94], which have been linked to behavioral dysregulation, cognitive impairment, and synaptic dysfunction [54, 95]. These immune alterations are closely linked to oxidative stress indicators—including elevated malondialdehyde (MDA) and decreased antioxidant enzyme activity (sod, gpz, gst)—which contribute to progressive neuronal damage [96–98]. Rather than detailing all transcriptomic and neurotrophic changes here, we refer the reader to Sections 5.1 and 6.1, where multi-omic analyses (e.g., immediate early gene (IEG), *bdnf* expression) and pharmacological rescue are discussed in depth as part of integrated therapeutic evaluation frameworks.

3.2 Behavioral and genotype-dependent evaluation of therapeutic responses in zebrafish

Zebrafish offer a scalable and genetically tractable platform for assessing the behavioral and molecular impact of neuroactive compounds across a variety of neuropsychiatric conditions, including anxiety, epilepsy, ASD, and schizophrenia. Standardized assays—such as the Novel Tank Diving Test (NTDT) for anxiety-like behavior, the Light-Dark Preference Test for stress responses, Social Preference and Shoaling Assays for social motivation, and Startle Response or Habituation paradigms for sensorimotor gating—have been validated across developmental stages and are widely used in both genetic and pharmacological models [11, 49, 79, 99]. These assays yield robust, dose-dependent behavioral phenotypes. For instance, PTZ induces convulsive activity reversed by valproate or carbamazepine in seizure models [100, 101], while embryonic VPA exposure elicits ASD-like traits ameliorated by risperidone or aripiprazole [79, 99]. Importantly, zebrafish behavioral phenotyping can be complemented with molecular and imaging approaches, including analysis of immediate early gene expression, neurotrophic markers (e.g., *bdnf*), and stress-related hormones, providing integrative readouts of drug efficacy. Reporter lines and transcriptomic profiling further strengthen translational relevance by enabling real-time monitoring of neuronal or immune activity *in vivo* (see also Section 5.1). A summary of major disease models, drug classes, and associated phenotypes is provided in **Table 3**.

Zebrafish also allow the dissection of genotype-specific pharmacological responses, critical for precision neuropsychopharmacology. CRISPR/Cas9-generated mutants for psychiatric risk genes such as *cacna1c*, *shank3a*, *syngap1*, *cntnap2*, and *nr3c2* display divergent behavioral phenotypes—including altered habituation, social deficits, and anxiety-like behavior—and respond differentially to SSRIs, antipsychotics, or modulators of glutamatergic and oxytocinergic signaling [64–66, 79]. For example, Banono et al. showed that only one allele of *cacna1c* responded to risperidone, illustrating allele-specific therapeutic effects [71]. Likewise, *nr3c2* mutants with altered glucocorticoid signaling retain responsiveness to fluoxetine [8]. Copy number variant (CNV) models, such as manipulations of the 16p11.2 locus (e.g., *kctd13*, *sez6l2*), reproduce human phenotypes, including micro- and macrocephaly, and exhibit partial rescue upon treatment with valproic acid or *bdnf* mimetics [102–104]. The development of mutant panels and the integration of behavioral assays with

Application	Model type	Drug classes tested	Key readouts	Ref.
ASD pharmacology	VPA exposure, CRISPR mutants	Antipsychotics, SSRIs	Social interaction, locomotion	[47, 49, 79, 84, 99]
Schizophrenia	*cacna1c*, *akt3* mutants	Antipsychotics	PPI, locomotion	[72, 73]
Alzheimer's/ neurodegeneration	OKA, Aβ overexpression	Histone deacetylase (HDAC) inhibitors, antioxidants	Memory tasks, tau/ Aβ pathology	[83, 91]
Substance abuse	Wild-type, drug-exposed larvae	Alcohol, nicotine, opioids	CPP, withdrawal, sensitization	[86]
General CNS screening	Wild-type or disease models	CNS-active libraries	Behavioral profiling, neurotoxicity	[10, 11]

Table 3.
Therapeutic screening applications in zebrafish.

transcriptomic analysis enable rapid stratification of responders and nonresponders across genetic backgrounds. These findings, increasingly aligned with human genome-wide association studies (GWAS) and polygenic risk score (PRS) data, support the application of zebrafish as a powerful system for early-stage pharmacogenomic screening and therapeutic discovery.

4. High-throughput screening and drug evaluation in zebrafish

Zebrafish have become a cornerstone of modern drug discovery, for all the advantages already mentioned. These features have facilitated the development of high-throughput screening (HTS) platforms that allow rapid evaluation of therapeutic candidates across a wide range of disease models, including neurological and neurodevelopmental disorders as well as cancer. The incorporation of automated technologies has further enhanced the scalability and precision of zebrafish-based assays, revolutionizing early-phase drug development workflows.

4.1 Automated systems for drug screening

Recent advances in HTS platforms have established zebrafish as a key model for early-stage drug discovery. Modern systems combine robotics, microfluidics, and high-content imaging to process thousands of larvae daily. For example, the WiScan Hermes platform automates brightfield and fluorescence imaging, enabling the quantification of complex cellular phenotypes such as GFP-labeled hematopoietic stem cells through AI-based analysis and customizable software [105]. Microfluidic technologies have also expanded the capabilities of organ-specific screening by enabling dynamic concentration gradients and real-time imaging. A notable example is the Automated *In Vivo* Screening System (AISS) developed at Sun Yat-sen University, which encapsulates zebrafish larvae in nanoliter droplets without anesthesia, preserving physiological integrity and allowing repeated measurements of cardiotoxicity, liver function, or neuronal activity in the same individual [106–108]. This system has proven effective in identifying drug candidates capable of rescuing epilepsy-like phenotypes in combination with behavioral assays and F0 CRISPR screening and has also revealed dose-dependent bradycardia induced by antipsychotics such as sertindole [108].

In parallel, the implementation of F0 knockout strategies allows rapid gene function analysis without the need to generate stable mutant lines. Microinjection of *gRNA*/ Cas9 complexes into one-cell-stage embryos yields observable phenotypes within days, reducing target validation timelines to approximately 3 months [109]. This facilitates early-stage drug-gene interaction studies with functional readouts. Automated platforms further enhance screening precision. WiSoft Athena, for instance, improves tracking accuracy by up to 31.6% in challenging visual conditions and supports simultaneous imaging of multiple cell populations, including myeloid and apoptotic markers [105, 110]. Fluorescence-based readouts also enable multiparametric assessments across hematopoietic, ocular, and apoptotic pathways. Altogether, these technological developments make zebrafish a scalable and cost-efficient screening model that bridges *in vitro* testing with mammalian systems while maintaining high biological relevance.

4.2 Toxicity assessment and LD50 determination

In addition to their role in disease modeling and drug efficacy evaluation, zebrafish are widely used for toxicity assessment across developmental stages. Acute toxicity is typically evaluated in embryos (0–96 hpf) using the fish embryo acute toxicity (FET) test, standardized by the Organization for Economic Co-operation and Development, which monitors four lethal endpoints: coagulation, absence of somite formation, non-detachment of the tail, and lack of heartbeat [111, 112]. Embryos exhibit reduced metabolic activity, minimizing false positives associated with inactive prodrugs [113]. In contrast, larvae (>72 hpf) show increased sensitivity due to functional organ systems and active xenobiotic metabolism. A meta-analysis of 600 chemicals revealed that larval LC_{50} values are, on average, 2.5 times lower than those of embryos, particularly for neurotoxic agents [114]. Beyond general assessments of mortality and gross malformations, specialized assays enable the evaluation of reproductive toxicity, hepatotoxicity, nephrotoxicity, cardiotoxicity, and liability to seizures or drug abuse [115, 116]. The zebrafish embryo developmental toxicity assay (ZEDTA) extends this approach by detecting sublethal effects such as pericardial edema, skeletal defects, and organ-specific toxicity, often using fluorescent transgenic lines for real-time visualization of affected tissues [112, 117, 118]. Standardized LD_{50} protocols involve 96-hour exposures to logarithmic drug concentrations, automated imaging of lethal and sublethal endpoints, and use of logD to predict the compound distribution and tissue accumulation.

Sublethal doses for efficacy studies are selected based on several criteria [117, 118]:

- the No Observed Adverse Effect Level (NOAEL), defined as the highest concentration without mortality or malformations,
- the oxidative stress threshold, i.e., doses below those causing significant reactive oxygen species (ROS) elevation (e.g., <2.5 μM E804), and
- behavioral viability, meaning concentrations that permit ≥80% normal swimming patterns.

Genetic zebrafish models allow mechanistic exploration of toxic responses through comparisons between wild-type and mutant lines. For example, *mrp1* knockouts exhibit impaired efflux capacity and increased sensitivity to chemotherapeutics, while *ahr2* mutants are resistant to TCDD-induced cardiotoxicity due

Feature	Wild-type applications	Mutant applications	Key findings	Ref.
Toxicokinetic	Baseline absorption/ distribution	*mrp1* mutants study efflux defects	Defects in toxicant efflux; increased sensitivity to drugs and xenobiotics	[119]
Organ vulnerability	General toxicity screening	*ahr2* mutants for cardiotoxicity	Mutants resistant to TCDD-induced cardiovascular toxicity	[120, 122]
		nrf2a mutants	Markedly higher mortality and malformations after Cd and Mn exposure	[121]
		gstp1 mutants	Increased MDA and ROS levels after toxicant exposure	[123]
Long-term effects	Acute toxicity profiling	Spinster heterozygotes model accelerated aging	Model of premature senescence and reduced lifespan	[95]

Table 4.
Genetic models enhance mechanistic toxicity studies.

to disruption of the aryl hydrocarbon receptor pathway [119, 120]. Similarly, *nrf2a* mutants display increased susceptibility to cadmium and manganese toxicity through heightened oxidative stress, and *gstp1* knockouts show exacerbated redox imbalance and endoplasmic reticulum stress under drug exposure [114, 115]. These tools also support the evaluation of organ-specific and long-term effects, given that compound uptake and clearance depend on developmental stage and transporter expression [121]. A summary of these applications is provided in **Table 4**, which highlights how specific mutant lines contribute to understanding toxicokinetics, organ vulnerability, and long-term consequences under defined exposure conditions. Overall, zebrafish genetic models contribute to a more personalized and predictive toxicology pipeline, enabling the identification of genotype-specific vulnerabilities and improving preclinical drug safety evaluation [95, 117–119].

5. Multi-omic approaches to neuropsychiatric drug evaluation

The growing implementation of multi-omic approaches in zebrafish neuropharmacology has enabled a shift from descriptive behavioral phenotyping toward mechanistically informed and predictive modeling of drug action. By integrating transcriptomic and proteomic data with behavioral readouts and network-based analysis, these approaches allow for the identification of molecular signatures underlying therapeutic efficacy, adverse effects, and genotype-specific responses. In contrast to earlier sections focusing on behavioral and molecular correlations, this section emphasizes how multi-omic data contribute to systems-level pharmacology, biomarker discovery, and translational alignment with human psychiatric genomics.

5.1 Network-based transcriptomic and proteomic profiling

High-throughput transcriptomic studies in zebrafish have revealed that psychotropic compounds such as fluoxetine, amphetamine, methylphenidate, and atomoxetine induce widespread and drug-specific gene expression changes. These changes span synaptic plasticity, lipid metabolism, neuroinflammatory signaling, circadian regulation, and apoptosis [124, 125]. Notably, methylphenidate and atomoxetine, both used clinically for attention deficit disorders, modulate cholesterol biosynthesis genes in opposing directions, correlating with divergent effects on anxiety-like behavior [125]. Fluoxetine, meanwhile, reverses chronic stress-induced transcriptional dysregulation, restoring behavioral and molecular homeostasis [126].

These transcriptomic shifts converge on functionally coherent gene networks. Weighted Gene Co-expression Network Analysis (WGCNA) has been used to detect modules enriched in serotonergic signaling, neurogenesis, and immune-related pathways in zebrafish exposed to stress and addictive substances [127]. These modules can be annotated using Gene Ontology (GO) or KEGG pathways and offer a structured framework for pharmacological classification and compound stratification.

Importantly, zebrafish modules are increasingly being compared to human psychiatric genomic datasets. Al-Soufi et al. demonstrated a cross-species alignment between zebrafish transcriptomic modules derived from nicotine and ethanol withdrawal and polygenic risk scores (PRS) from human cohorts with alcohol dependence, identifying conserved transcriptional programs involving synaptic regulation and neuroimmune pathways [127]. This supports the translational validity of zebrafish multi-omic data for modeling human neuropsychiatric vulnerability.

Proteomic profiling adds complementary insight by capturing activity-dependent signaling cascades, protein-protein interactions, and posttranslational modifications not evident at the transcript level. For example, phosphoproteomic and secretomic analyses in zebrafish have revealed drug-induced changes in synaptic vesicle trafficking, kinase activation, and neurotrophic signaling [125, 127]. Recent advances in LC-MS/MS have enabled regional and cell-type-specific profiling, offering greater spatial and functional resolution of drug effects.

Although less commonly used, metabolomic and epigenomic profiling are also emerging in zebrafish studies, providing additional layers of information on drug-induced shifts in cellular energetics, redox balance, or chromatin remodeling. These layers are especially promising for identifying long-term effects and persistent molecular imprints following early-life exposures.

Table 5 highlights representative studies where behavioral phenotypes correlate with specific molecular responses, supporting the development of multiparametric models for therapeutic screening.

In addition to drug exposure models, CRISPR-based genetic manipulations have facilitated the identification of genotype-specific molecular responses. In *shank3a* mutant zebrafish, a model for ASD, transcriptomic analysis revealed altered synaptic signaling networks that were partially rescued by risperidone, linking pharmacological and genetic modulation [125].

Emerging computational tools—including dimensionality reduction, clustering, and supervised learning—are now being applied to zebrafish omic datasets to classify compounds, identify responder subpopulations, and predict behavioral outcomes based on molecular profiles [127]. While their use in zebrafish neuropharmacology remains limited compared to mammalian systems, these methods are expected to become increasingly important as single-cell transcriptomics, spatial omics, and high-resolution imaging become standard.

Compound/ treatment	Zebrafish model	Molecular response	Behavioral response	Ref.
Fluoxetine	Adult, CUS (chronic unpredictable stress)	Reversal of CUS-induced gene expression changes in neuroinflammatory and synaptic remodeling pathways	Rescue of anxiety-like behavior	[128]
Amphetamine	Larvae, developmental exposure	Downregulation of immediate early genes (*egr1*, *fosab*, *junbb*); disruptions in apoptosis, circadian rhythm, immune signaling	Impaired locomotion, and reduced habituation to acoustic stimuli	[104]
Nicotine				
Oxycodone				
Methylphenidate	Adult, 8-day exposure	Upregulation of cholesterol biosynthesis genes (*hmgcs1*, *hmgcra*, *fdps*, *sqlea*)	Increased anxiety-like behavior (less time in the top zone)	[129]
Atomoxetine	Adult, 8-day exposure	Downregulation of the same cholesterol biosynthesis genes	Reduced anxiety-like behavior (more time in the top zone)	[129]

Table 5.
Summary of key studies linking transcriptomics and behavior.

6. Future perspectives and clinical relevance

Zebrafish models have emerged as a central platform for translational neuropsychiatry, bridging mechanistic neuroscience and therapeutic development. Their capacity to integrate behavioral phenotyping, high-throughput pharmacological screening, and real-time molecular imaging in a whole-organism system makes them especially valuable for identifying conserved neural mechanisms and testing drug responses. Building on the insights presented in previous sections, this chapter outlines the translational relevance of zebrafish, discusses current model limitations, and highlights future directions toward precision psychiatry.

6.1 From mechanistic models to precision psychiatry

Zebrafish have transitioned from serving as descriptive models of psychiatric phenotypes to becoming powerful mechanistic tools that support the development of precision psychiatry. Their conserved neuroanatomy—including the pallium, diencephalon, and cerebellum—allows the modeling of core behavioral and molecular traits relevant to disorders such as anxiety, schizophrenia, ASD, and neurodegeneration [78, 82, 84]. A major strength lies in their compatibility with dual-readout strategies, where behavioral phenotyping can be coupled with molecular profiling in the same individual (as described in Section 4.2). For example, VPA-induced models of ASD not only recapitulate social and locomotor impairments but also respond to antipsychotics like risperidone and aripiprazole with parallel restoration of IEG activity and suppression of inflammatory markers [64, 101, 102, 128]. IEGs such as*fosab*, *egr1*, *egr4*, and *junbb* act as sensitive and rapid transcriptional markers of neuronal activation, and their regulation reflects key processes including synaptic plasticity, stress reactivity, apoptosis, and immune signaling [92, 129]. Among them, *fosab* has emerged as a particularly robust indicator of neuronal excitation in response to both acute and chronic stimuli. Similarly, in Alzheimer's models, compounds such as curcumin or HDAC inhibitors mitigate tau and Aβ pathology while improving *bdnf* expression and cognitive performance [11, 91]. These findings reflect a convergence between phenotype and molecular response that reinforces the relevance of zebrafish for early-phase pharmacodynamic validation and biomarker-guided drug screening.

Importantly, CRISPR-based genetic models reveal allele-specific and gene-dependent variations in drug sensitivity. For instance, distinct *cacna1c* mutant lines display divergent behavioral phenotypes and differential responses to risperidone, highlighting the need for allelic stratification in pharmacogenomic research [71]. Likewise, mutations in ASD-linked genes such as *shank3a* and *syngap1* lead to non-equivalent drug responses, emphasizing the need to tailor therapies based on genetic background [83]. Emerging tools such as transgenic biosensors (e.g., GCaMP, NF-κB) and CRISPR panels developed by Capps et al. allow real-time monitoring of neural and immune activity and multiplexed screening across ASD-associated loci [74]. The integration of zebrafish multi-omic data with human polygenic risk scores, especially in depression and addiction models, further enhances translational value and supports personalized therapeutic strategies [74, 127]. Altogether, zebrafish provide a uniquely scalable and genetically tractable model for bridging experimental neuroscience and individualized medicine, enabling the discovery and validation of targeted interventions grounded in mechanistic insight.

6.2 Limitations and model constraints

Despite their many advantages, zebrafish models present several biological and technical limitations that must be considered. A key challenge is drug administration: while immersion remains the standard route for compound delivery due to its simplicity and scalability, it offers limited control over internal concentrations. Passive absorption through the skin and gills can lead to interindividual variability and hinder precise pharmacokinetic profiling, complicating dose-response analyses [8, 105]. In adult zebrafish, procedural constraints limit the use of certain invasive techniques, such as intracranial injections or electrophysiology, which are more readily applied in mammalian systems [130]. Moreover, while larval stages are highly suitable for imaging and genetic manipulation, they may not capture late-onset behavioral phenotypes or complex neuropsychiatric traits like hallucinations or language impairment, which are uniquely human and difficult to model in lower vertebrates [8, 9].

Another limitation stems from the teleost-specific whole genome duplication, which increases genetic redundancy. Many genes exist in duplicated pairs (ohnologs), which can have overlapping or divergent functions, potentially buffering the effects of single-gene mutations or confounding transcriptomic and proteomic analyses [8, 74]. This complexity necessitates rigorous validation of gene-editing targets and careful interpretation of functional assays, especially in CRISPR-based models of neuropsychiatric disorders. Nonetheless, these challenges do not negate the zebrafish's value. Rather, they highlight the importance of using zebrafish as a complementary tool within a broader translational pipeline. Their rapid development, genetic tractability, and whole-organism readouts position zebrafish as an efficient, cost-effective model for hypothesis generation and early-stage therapeutic screening that can inform and refine subsequent mammalian studies [8, 10, 105].

7. Conclusions

Zebrafish have emerged as a powerful and versatile model for neuropsychiatric drug discovery, uniquely combining genetic manipulability, behavioral scalability, and compatibility with high-throughput screening platforms. As demonstrated throughout this chapter, zebrafish support the simultaneous assessment of behavioral, molecular, and genetic outcomes in a whole-organism context. This integrative capacity is especially valuable for evaluating therapeutic candidates in polygenic and heterogeneous disorders such as ASD, schizophrenia, and addiction. CRISPR-based disease models (such as *shank3a*, *cacna1c*, or *nr3c2*), pharmacological paradigms (including PTZ or valproic acid), and transcriptomic profiling of markers like *bdnf* or immediate early genes collectively enable the mapping of drug effects from neural circuit activity to gene expression. These tools foster the identification of biomarker-phenotype convergence, a key step toward mechanism-based precision pharmacology.

Despite their advantages, zebrafish models face certain challenges. Their divergent brain anatomy and exposure routes may limit the extrapolation of some human-specific features, and gene duplication events in the zebrafish genome can complicate genetic interpretation. Nonetheless, rapid technological innovations—including spatial transcriptomics, patient-specific gene editing, and AI-driven behavioral analysis—are expanding the scope and precision of zebrafish-based neuroscience. Rather than serving as a replacement for rodent or clinical studies, zebrafish complement existing models by offering early-stage, mechanistic insight, and scalable screening capacity. As these tools mature, zebrafish will continue to bridge basic research and translational psychiatry, contributing to the development of more effective and personalized neuropsychiatric care.

Abbreviations

AD	Alzheimer's disease
ALS	amyotrophic lateral sclerosis
ASD	autism spectrum disorder
bdnf	brain-derived neurotrophic factor
CNV	copy number variant
CNS	central nervous system
CPP	conditioned place preference
CRISPR	clustered regularly interspaced short palindromic repeats
FASD	fetal alcohol spectrum disorder
FET	fish embryo acute toxicity
GFP	green fluorescent protein
GPx	glutathione peroxidase
GSK3β	glycogen synthase kinase 3 beta
GST	glutathione S-transferase
HDAC	histone deacetylase
HTS	high-throughput screening
IEG	immediate early gene
LC_{50}	lethal concentration 50
LD_{50}	lethal dose 50
MDA	malondialdehyde
MO	morpholino antisense oligonucleotide
NOAEL	no observed adverse effect level
OKA	okadaic acid
PD	Parkinson's disease

PPI	prepulse inhibition
PRS	polygenic risk score
PTZ	pentylenetetrazole
ROS	reactive oxygen species
SOD	superoxide dismutase
SSRI	selective serotonin reuptake inhibitor
VPA	valproic acid
WT	wild-type
ZEDTA	zebrafish embryo developmental toxicity assay

Author details

Antonella Damiano[1], Laura Sánchez[2] and Álvaro Arana[2]*

1 Department of Bioscience and Technology for Food, Agriculture and Environment, University of Teramo, Teramo, Italy

2 Department of Zoology, Genetics and Physical Anthropology, Faculty of Veterinary Medicine, University of Santiago de Compostela, Lugo, Spain

*Address all correspondence to: alvaro.arana@usc.es

References

[1] Wullimann MF, Rupp B, Reichert H. Neuroanatomy of the Zebrafish Brain. Basel: Birkhäuser Basel; 1996

[2] Kimmel CB. Patterning the brain of the zebrafish embryo. Annual Review of Neuroscience. 1993;**16**:707-732

[3] Bae YK, Kani S, Shimizu T, Tanabe K, Nojima H, Kimura Y, et al. Anatomy of zebrafish cerebellum and screen for mutations affecting its development. Developmental Biology. 2009;**330**(2):406-426

[4] Wullimann MF, Mueller T. Teleostean and mammalian forebrains contrasted: Evidence from genes to behavior. Journal of Comparative Neurology. 2004;**475**(2):143-162

[5] Wullimann MF. Secondary neurogenesis and telencephalic organization in zebrafish and mice: A brief review. Integrative Zoology. 2009;**4**(1):123-133

[6] Chapouton P, Godinho L. Neurogenesis. In: Methods in Cell Biology. Elsevier; 2010. pp. 72-126

[7] Schmidt R, Strähle U, Scholpp S. Neurogenesis in zebrafish—From embryo to adult. Neural Development. 2013;**8**(1):3

[8] Fontana BD, Mezzomo NJ, Kalueff AV, Rosemberg DB. The developing utility of zebrafish models of neurological and neuropsychiatric disorders: A critical review. Experimental Neurology. 2018;**299**(Pt A):157-171. DOI: 10.1016/j.expneurol.2017.10.004

[9] Stewart AM, Braubach O, Spitsbergen J, Gerlai R, Kalueff AV. Zebrafish models for translational neuroscience research: From tank to bedside. Trends in Neurosciences. 2014;**37**(5):264-278

[10] Kalueff AV, Stewart AM, Gerlai R. Zebrafish as an emerging model for studying complex brain disorders. Trends in Pharmacological Sciences. 2014;**35**(2):63-75

[11] Khan KM, Collier AD, Meshalkina DA, Kysil EV, Khatsko SL, Kolesnikova T, et al. Zebrafish models in neuropsychopharmacology and CNS drug discovery. British Journal of Pharmacology. 2017;**174**(13):1925-1944

[12] Bashirzade AA, Zabegalov KN, Volgin AD, Belova AS, Demin KA, de Abreu MS, et al. Modeling neurodegenerative disorders in zebrafish. Neuroscience and Biobehavioral Reviews. 2022;**138**:104679

[13] Grossman L, Stewart A, Utterback E, Wu N, Dileo J, Frank K, et al. Effects of piracetam on behavior and memory in adult zebrafish. Brain Research Bulletin. 2011;**85**:58-63

[14] Bailey J, Oliveri A, Levin ED. Zebrafish model systems for developmental neurobehavioral toxicology. Birth Defects Research. Part C, Embryo Today. 2013;**99**(1):14-23

[15] Cheresiz SV, Volgin AD, Kokorina Evsyukova A, Bashirzade AAO, Demin KA, de Abreu MS, et al. Understanding neurobehavioral genetics of zebrafish. Journal of Neurogenetics. 2020;**34**(2):203-215

[16] Michael Stewart A, Kalueff AV. The developing utility of zebrafish models for cognitive enhancers research. Current Neuropharmacology. 2012;**10**(3):263-271

[17] Kalueff AV, Gebhardt M, Stewart AM, Cachat JM, Brimmer M, Chawla JS, et al. Towards a comprehensive catalog of zebrafish behavior 1.0 and beyond. Zebrafish. 2013;**10**(1):70-86

[18] Levin ED, Cerutti DT. Behavioral neuroscience of zebrafish. In: Buccafusco JJ, editor. Methods of Behavior Analysis in Neuroscience, Frontiers in Neuroscience. 2nd ed. Boca Raton (FL): CRC Press/Taylor & Francis; 2009

[19] Rink E, Wullimann MF. Development of the catecholaminergic system in the early zebrafish brain: An immunohistochemical study. Brain Research. Developmental Brain Research. 2002;**137**(1):89-100

[20] Ma PM, Lopez M. Consistency in the number of dopaminergic paraventricular organ-accompanying neurons in the posterior tuberculum of the zebrafish brain. Brain Research. 2003;**967**(1-2):267-272

[21] Ryu S, Holzschuh J, Mahler J, Driever W. Genetic analysis of dopaminergic system development in zebrafish. In: Riederer P, Reichmann H, Youdim MBH, Gerlach M, editors. Parkinson's Disease and Related Disorders. Vienna: Springer Vienna; 2006. pp. 61-66

[22] Bretaud S, Lee S, Guo S. Sensitivity of zebrafish to environmental toxins implicated in Parkinson's disease. Neurotoxicology and Teratology. 2004;**26**(6):857-864

[23] Lam CS, Korzh V, Strahle U. Zebrafish embryos are susceptible to the dopaminergic neurotoxin MPTP. European Journal of Neuroscience. 2005;**21**(6):1758-1762

[24] McKinley ET, Baranowski TC, Blavo DO, Cato C, Doan TN, Rubinstein AL. Neuroprotection of MPTP-induced toxicity in zebrafish dopaminergic neurons. Molecular Brain Research. 2005;**141**(2):128-137

[25] Sallinen V, Torkko V, Sundvik M, Reenilä I, Khrustalyov D, Kaslin J, et al. MPTP and MPP+ target specific aminergic cell populations in larval zebrafish. Journal of Neurochemistry. 2009;**108**(3):719-731

[26] Giacomini NJ, Rose B, Kobayashi K, Guo S. Antipsychotics produce locomotor impairment in larval zebrafish. Neurotoxicology and Teratology. 2006;**28**(2):245-250

[27] Boehmler W, Carr T, Thisse C, Thisse B, Canfield VA, Levenson R. D4 dopamine receptor genes of zebrafish and effects of the antipsychotic clozapine on larval swimming behaviour. Genes, Brain, and Behavior. 2007;**6**(2):155-166

[28] Anichtchik OV, Kaslin J, Peitsaro N, Scheinin M, Panula P. Neurochemical and behavioural changes in zebrafish Danio rerio after systemic administration of 6-hydroxydopamine and 1-methyl-4-phenyl-1,2,3,6-tetrahydropyridine. Journal of Neurochemistry. 2004;**88**(2):443-453

[29] Kirby BB, Takada N, Latimer AJ, Shin J, Carney TJ, Kelsh RN, et al. In vivo time-lapse imaging shows dynamic oligodendrocyte progenitor behavior during zebrafish development. Nature Neuroscience. 2006;**9**(12):1506-1511

[30] Kawai H, Arata N, Nakayasu H. Three-dimensional distribution of astrocytes in zebrafish spinal cord. Glia. 2001;**36**(3):406-413

[31] Peri F, Nüsslein-Volhard C. Live imaging of neuronal degradation by microglia reveals a role for v0-ATPase

a1 in phagosomal fusion In vivo. Cell. 2008;**133**(5):916-927

[32] Vila M, Jackson-Lewis V, Guégan C, Chu WD, Teismann P, Choi DK, et al. The role of glial cells in Parkinson's disease. Current Opinion in Neurology. 2001;**14**(4):483

[33] Mullett SJ, Hinkle DA. DJ-1 knock-down in astrocytes impairs astrocyte-mediated neuroprotection against rotenone. Neurobiology of Disease. 2009;**33**(1):28-36

[34] Jeong JY, Kwon HB, Ahn JC, Kang D, Kwon SH, Park JA, et al. Functional and developmental analysis of the blood–brain barrier in zebrafish. Brain Research Bulletin. 2008;**75**(5):619-628

[35] Najib NHM, Nies YH, Halim SASA, Yahaya MF, Das S, Lim WL, et al. Modeling Parkinson's Disease in Zebrafish. CNS & Neurological Disorders - Drug Targets. 2025;**19**(5):386-399

[36] Doyle JM, Croll RP. A critical review of zebrafish models of Parkinson's disease. Frontiers in Pharmacology. 2022;**13**:835827. DOI: 10.3389/fphar.2022.835827

[37] Wadan AHS, Mohamed W. Chapter 13—Various zebrafish models of Parkinson's disease: What gives us hope. In: Mohamed W, editor. Translational Models of Parkinson's Disease and Related Movement Disorders. London: Academic Press; 2025. pp. 219-230

[38] Newman M, Ebrahimie E, Lardelli M. Using the zebrafish model for Alzheimer's disease research. Frontiers in Genetics. 2014;**5**:189. DOI: 10.3389/fgene.2014.00189

[39] Chen M, Martins RN, Lardelli M. Complex splicing and neural expression of duplicated tau genes in zebrafish embryos. Journal of Alzheimer's Disease. 2009;**18**(2):305-317

[40] Kumar V, Singh C, Singh A. Zebrafish an experimental model of Huntington's disease: Molecular aspects, therapeutic targets and current challenges. Molecular Biology Reports. 2021;**48**(12):8181-8194

[41] D'Amora M, Galgani A, Marchese M, Tantussi F, Faraguna U, De Angelis F, et al. Zebrafish as an innovative tool for epilepsy modeling: State of the art and potential future directions. International Journal of Molecular Sciences. 2023;**24**(9):7702

[42] LaCoursiere CM, Ullmann JFP, Koh HY, Turner L, Baker CM, Robens B, et al. Zebrafish models of candidate human epilepsy-associated genes provide evidence of hyperexcitability. bioRxiv [Preprint]. 2024:2024.02.07.579190. DOI: 10.1101/2024.02.07.579190. Update in: iScience. 2024 Jun 05;27(7):110172. DOI: 10.1016/j.isci.2024.110172

[43] Borba JV, Canzian J, Resmim CM, Silva RM, Duarte MCF, Mohammed KA, et al. Towards zebrafish models to unravel translational insights of obsessive-compulsive disorder: A neurobehavioral perspective. Neuroscience and Biobehavioral Reviews. 2024;**162**:105715

[44] Ramesh T, Lyon AN, Pineda RH, Wang C, Janssen PML, Canan BD, et al. A genetic model of amyotrophic lateral sclerosis in zebrafish displays phenotypic hallmarks of motoneuron disease. Disease Models & Mechanisms. 2010;**3**(9-10):652-662

[45] Ciura S, Lattante S, Le Ber I, Latouche M, Tostivint H, Brice A, et al. Loss of function of C9orf72 causes motor deficits in a zebrafish model of

amyotrophic lateral sclerosis. Annals of Neurology. 2013;**74**(2):180-187

[46] McGown A, McDearmid JR, Panagiotaki N, Tong H, Al Mashhadi S, Redhead N, et al. Early interneuron dysfunction in ALS: Insights from a mutant sod1 zebrafish model. Annals of Neurology. 2013;**73**(2):246-258

[47] Stewart AM, Nguyen M, Wong K, Poudel MK, Kalueff AV. Developing zebrafish models of autism spectrum disorder (ASD). Progress in Neuro-Psychopharmacology & Biological Psychiatry. 2014;**50**:27-36

[48] Al-Zoubi RM, Abu-Hijleh H, Zarour A, Zakaria ZZ, Yassin A, Al-Ansari AA, et al. Zebrafish model in illuminating the complexities of post-traumatic stress disorders: A unique research tool. International Journal of Molecular Sciences. 2024;**25**(9):4895

[49] Meshalkina DA, Kizlyk MN, Kysil EV, Collier AD, Echevarria DJ, Abreu MS, et al. Zebrafish models of autism spectrum disorder. Experimental Neurology. 2018;**299**(Pt A):207-216

[50] Ziv L, Muto A, Schoonheim PJ, Meijsing SH, Strasser D, Ingraham HA, et al. An affective disorder in zebrafish with mutation of the glucocorticoid receptor. Molecular Psychiatry. 2013;**18**(6):681-691

[51] Griffiths BB, Schoonheim PJ, Ziv L, Voelker L, Baier H, Gahtan E. A zebrafish model of glucocorticoid resistance shows serotonergic modulation of the stress response. Frontiers in Behavioral Neuroscience. 2012;**6**:68

[52] Swain HA, Sigstad C, Scalzo FM. Effects of dizocilpine (MK-801) on circling behavior, swimming activity, and place preference in zebrafish (*Danio rerio*). Neurotoxicology and Teratology. 2004;**26**(6):725-729

[53] Best JD, Alderton WK. Zebrafish: An in vivo model for the study of neurological diseases. Neuropsychiatric Disease and Treatment. 2008;**4**(3):567-576

[54] Gerlai R. Zebrafish and relational memory: Could a simple fish be useful for the analysis of biological mechanisms of complex vertebrate learning? Behavioural Processes. 2017;**141**:242-250

[55] Barlow IL, Rihel J. Zebrafish sleep: From geneZZZ to neuronZZZ. Current Opinion in Neurobiology. 2017;**44**:65-71

[56] Müller UK, Van Leeuwen JL. Swimming of larval zebrafish: Ontogeny of body waves and implications for locomotory development. The Journal of Experimental Biology. 2004;**207**(5):853-868

[57] Kim DH, Kim J, Marques JC, Grama A, Hildebrand DGC, Gu W, et al. Pan-neuronal calcium imaging with cellular resolution in freely swimming zebrafish. Nature Methods. 2017;**14**(11):1107-1114

[58] Hubbard JM, Böhm UL, Prendergast A, Tseng PEB, Newman M, Stokes C, et al. Intraspinal sensory neurons provide powerful inhibition to motor circuits ensuring postural control during locomotion. Current Biology. 2016;**26**(21):2841-2853

[59] Kim YH, Lee Y, Kim D, Jung MW, Lee CJ. Scopolamine-induced learning impairment reversed by physostigmine in zebrafish. Neuroscience Research. 2010;**67**(2):156-161

[60] Luo L, Yan T, Yang L, Zhao M. Aluminum chloride and D-galactose induced a zebrafish model of

Alzheimer's disease with cognitive deficits and aging. Computational and Structural Biotechnology Journal. 2024;**23**:2230-2239

[61] Karimi Z, Zarifkar A, Dianatpour M, Mirzaei E, Dara M, Aligholi H. Finding a proper valproic acid-based autism spectrum disorder model in zebrafish: Early and long-term neurobehavioral studies. Iranian Journal of Psychiatry and Behavioral Sciences. 2023;**17**(3)e137118. Available from: https://brieflands.com/articles/ijpbs-137118#abstract

[62] Camussi D, Marchese M, Nicoletti F, Santorelli FM, Ogi A. Valproate-induced model of autism in adult zebrafish: A systematic review. Cells. 2025;**14**(2):109

[63] Huang RQ, Bell-Horner CL, Dibas MI, Covey DF, Drewe JA, Dillon GH. Pentylenetetrazole-induced inhibition of recombinant γ-aminobutyric acid type A (GABAA) receptors: Mechanism and site of action. The Journal of Pharmacology and Experimental Therapeutics. 2001;**298**(3):986-995

[64] Gauthier J, Champagne N, Lafrenière RG, Xiong L, Spiegelman D, Brustein E, et al. De novo mutations in the gene encoding the synaptic scaffolding protein SHANK3 in patients ascertained for schizophrenia. Proceedings of the National Academy of Sciences of the United States of America. 2010;**107**(17):7863-7868

[65] Sakai C, Ijaz S, Hoffman EJ. Zebrafish models of neurodevelopmental disorders: Past, present, and future. Frontiers in Molecular Neuroscience. 2018;**11**:294

[66] Liu Q, Yuan Y, Zhu F, Hong Y, Ge R. Efficient genome editing using CRISPR/Cas9 ribonucleoprotein approach in cultured medaka fish cells. Biology Open. 10 Aug 2018;7(8):bio035170

[67] Oksenberg N, Stevison L, Wall JD, Ahituv N. Function and regulation of AUTS2, a gene implicated in autism and human evolution. PLoS Genetics. 2013;**9**(1):e1003221

[68] Patowary A, Won SY, Oh SJ, Nesbitt RR, Archer M, Nickerson D, et al. Family-based exome sequencing and case-control analysis implicate CEP41 as an ASD gene. Translational Psychiatry. 2019;**9**(1):4

[69] Kozol RA, Cukier HN, Zou B, Mayo V, De Rubeis S, Cai G, et al. Two knockdown models of the autism genes SYNGAP1 and SHANK3 in zebrafish produce similar behavioral phenotypes associated with embryonic disruptions of brain morphogenesis. Human Molecular Genetics. 2015;**24**(14):4006-4023

[70] Kozol RA, Dallman JE. Drugs prescribed for Phelan-McDermid syndrome differentially impact sensory behaviors in shank3 zebrafish models. F1000Research. 2023;**12**:84. DOI: 10.12688/f1000research.127830.2

[71] Banono NS et al. Functional characterisation of single nucleotide variants of the psychiatric risk gene *cacna1c* in the zebrafish. bioRxiv 2021.09.30.462600. DOI: 10.1101/2021.09.30.462600

[72] Banono NS, Gawel K, De Witte L, Esguerra CV. Zebrafish larvae carrying a splice variant mutation in cacna1d: A new model for schizophrenia-like behaviours? Molecular Neurobiology. 2021;**58**(2):877-894

[73] Tan JXM, Ang RJW, Wee CL. Larval zebrafish as a model for mechanistic discovery in mental health. Frontiers in Molecular Neuroscience. 2022;**15**:900213

[74] Capps MES et al. Disrupted diencephalon development and neuropeptidergic pathways in zebrafish with autism-risk mutations. Proceedings of the National Academy of Sciences of the United States of America. 2025;**122**(23):e2402557122

[75] Baronio D, Chen YC, Panula P. Abnormal brain development of monoamine oxidase mutant zebrafish and impaired social interaction of heterozygous fish. Disease Models and Mechanisms. 2022;**15**(3):dmm049133

[76] Grone BP, Marchese M, Hamling KR, Kumar MG, Krasniak CS, Sicca F, et al. Epilepsy, behavioral abnormalities, and physiological comorbidities in syntaxin-binding protein 1 (STXBP1) mutant zebrafish. PLoS ONE. 2016;**11**(3):e0151148

[77] Ribeiro D, Nunes AR, Teles M, Anbalagan S, Blechman J, Levkowitz G, et al. Genetic variation in the social environment affects behavioral phenotypes of oxytocin receptor mutants in zebrafish. eLife. 2020;**9**:e56973

[78] Habicher J, Sanvido I, Bühler A, Sartori S, Piccoli G, Carl M. The risk genes for neuropsychiatric disorders negr1 and opcml are expressed throughout zebrafish brain development. Genes. 2024;**15**(3):363

[79] Tayanloo-Beik A, Hamidpour SK, Abedi M, Shojaei H, Tavirani MR, Namazi N, et al. Zebrafish modeling of autism spectrum disorders, current status and future prospective. Frontiers in Psychiatry. 2022;**13**:911770

[80] Lovely CB, Swartz ME, McCarthy N, Norrie JL, Eberhart JK. Bmp signaling mediates endoderm pouch morphogenesis by regulating Fgf signaling in zebrafish. Development. 1 Jun 2016;**143**(11):2000-2011

[81] Fernandes Y, Lovely CB. Zebrafish models of fetal alcohol spectrum disorders. Genesis. 2021;**59**(11):e23460

[82] Kozol RA, Abrams AJ, James DM, Buglo E, Yan Q, Dallman JE. Function over form: Modeling groups of inherited neurological conditions in zebrafish. Frontiers in Molecular Neuroscience. 2016;**9**:55. DOI: 10.3389/fnmol.2016.00055

[83] Nada SE, Williams FE, Shah ZA. Development of a novel and robust pharmacological model of okadaic acid-induced Alzheimer's disease in zebrafish. CNS & Neurological Disorders Drug Targets. 2016;**15**(1):86-94

[84] Vaz AP, Inati SK, Brunel N, Zaghloul KA. Coupled ripple oscillations between the medial temporal lobe and neocortex retrieve human memory. Science. 2019;**363**(6430):975-978

[85] Panula P, Sallinen V, Sundvik M, Kolehmainen J, Torkko V, Tiittula A, et al. Modulatory neurotransmitter systems and behavior: Towards zebrafish models of neurodegenerative diseases. Zebrafish. 2006;**3**(2):235-247

[86] Müller TE, Fontana BD, Bertoncello KT, Franscescon F, Mezzomo NJ, Canzian J, et al. Understanding the neurobiological effects of drug abuse: Lessons from zebrafish models. Progress in Neuro-Psychopharmacology and Biological Psychiatry. 2020;**100**:109873

[87] Wee CL, Song E, Nikitchenko M, Herrera KJ, Wong S, Engert F, et al. Social isolation modulates appetite and avoidance behavior via a common oxytocinergic circuit in larval zebrafish. Nature Communications. 2022;**13**(1):2573

[88] Saleem S, Kannan RR. Zebrafish: An emerging real-time model system

to study Alzheimer's disease and neurospecific drug discovery. Cell Death Discovery. 2018;**4**(1):45

[89] Barthelson K, Pederson SM, Newman M, Jiang H, Lardelli M. In-frame and frameshift mutations in zebrafish presenilin 2 affect different cellular functions in young adult brains. Journal of Alzheimer's Disease Reports. 2021;**5**(1):395-404

[90] Koehler D, Shah ZA, Williams FE. The GSK3β inhibitor, TDZD-8, rescues cognition in a zebrafish model of okadaic acid-induced Alzheimer's disease. Neurochemistry International. 2019;**122**:31-37

[91] Chia K, Klingseisen A, Sieger D, Priller J. Zebrafish as a model organism for neurodegenerative disease. Frontiers in Molecular Neuroscience. 2022;**15**:940484

[92] Mojzesz M, Widziolek M, Adamek M, Orzechowska U, Podlasz P, Prajsnar TK, et al. Tilapia Lake virus-induced neuroinflammation in zebrafish: Microglia activation and sickness behavior. Frontiers in Immunology. 2021;**12**:760882. DOI: 10.3389/fimmu.2021.760882

[93] Calcia MA, Bonsall DR, Bloomfield PS, Selvaraj S, Barichello T, Howes OD. Stress and neuroinflammation: A systematic review of the effects of stress on microglia and the implications for mental illness. Psychopharmacology. 2016;**233**(9):1637-1650

[94] Reus LM, Jansen IE, Tijms BM, Visser PJ, Tesi N, van der Lee SJ, et al. Connecting dementia risk loci to the CSF proteome identifies pathophysiological leads for dementia. Brain;**147**(10):3522-3533

[95] Sasaki T, Kishi S. Molecular and chemical genetic approaches to developmental origins of aging and disease in zebrafish. Biochimica et Biophysica Acta (BBA) - Molecular Basis of Disease. 2013;**1832**(9):1362-1370

[96] Yang Y, Liu W, Mu X, Qi S, Fu B, Wang C. Biological response of zebrafish embryos after short-term exposure to thifluzamide. Scientific Reports. 2016;**6**(1):38485

[97] Huang X, Li Y, Wang T, Liu H, Shi J, Zhang X. Evaluation of the oxidative stress status in zebrafish (*Danio rerio*) liver induced by three typical organic UV filters (BP-4, PABA and PBSA). International Journal of Environmental Research and Public Health. 2020;**17**(2):651

[98] Arthi B, Chellathai D. Effects of vasicine in neuroinflammatory zebrafish model. Bioinformation. 2023;**19**(5):595-604

[99] Hara Y, Ago Y, Taruta A, Hasebe S, Kawase H, Tanabe W, et al. Risperidone and aripiprazole alleviate prenatal valproic acid-induced abnormalities in behaviors and dendritic spine density in mice. Psychopharmacology. 2017;**234**(21):3217-3228

[100] Baraban SC, Taylor MR, Castro PA, Baier H. Pentylenetetrazole induced changes in zebrafish behavior, neural activity and c-fos expression. Neuroscience. 2005;**131**(3):759-768

[101] Zhao SS, Wang YL, Sun MZ, Lu L, Wang YN, Pfister D, et al. Drug screening: Zebrafish as a tool for studying epileptic-related chemical compounds. Protein & Cell. 2015;**6**(11):853-857

[102] Gambini J, Stromsnes K. Oxidative stress and inflammation: From mechanisms to therapeutic approaches. Biomedicine. 2022;**10**(4):753

[103] Sahu MP, Pazos-Boubeta Y, Pajanoja C, Rozov S, Panula P, Castrén E. Neurotrophin receptor Ntrk2b function in the maintenance of dopamine and serotonin neurons in zebrafish. Scientific Reports. 2019;**9**(1):2036

[104] Barbereau C, Yehya A, Silhol M, Cubedo N, Verdier JM, Maurice T, et al. Neuroprotective brain-derived neurotrophic factor signaling in the TAU-P301L tauopathy zebrafish model. Pharmacological Research. 2020;**158**:104865

[105] Lubin A, Otterstrom J, Hoade Y, Bjedov I, Stead E, Whelan M, et al. A versatile, automated and high-throughput drug screening platform for zebrafish embryos. Biology Open. 2021;**10**(9):bio058513

[106] Candelier R, Murmu MS, Romano SA, Jouary A, Debrégeas G, Sumbre G. A microfluidic device to study neuronal and motor responses to acute chemical stimuli in zebrafish. Scientific Reports. 2015;**5**:12196

[107] Erickstad M, Hale LA, Chalasani SH, Groisman A. A microfluidic system for studying the behavior of zebrafish larvae under acute hypoxia. Lab on a Chip. 2015;**15**(3):857-866

[108] Duan J, Lin G, Jiao K, Hong X, Lin X. Fully automated in vivo screening system for multi-organ imaging and pharmaceutical evaluation. Microsystems & Nanoengineering. 2025;**11**(1):1-11

[109] Locubiche S, Ordóñez V, Abad E, Scotto di Mase M, Di Donato V, De Santis F. A zebrafish-based platform for high-throughput epilepsy modeling and drug screening in F0. International Journal of Molecular Sciences. 2024;**25**(5):2991

[110] Wang X, Cheng E, Burnett IS, Huang Y, Wlodkowic D. Automatic multiple zebrafish larvae tracking in unconstrained microscopic video conditions. Scientific Reports. 2017;**7**(1):17596

[111] OECD. Test No. 236: Fish Embryo Acute Toxicity (FET) Test. OECD; 2013

[112] Hoyberghs J, Bars C, Pype C, Foubert K, Ayuso Hernando M, Van Ginneken C, et al. Refinement of the zebrafish embryo developmental toxicity assay. MethodsX. 2020;**7**:101087

[113] Nawaji T, Mizoguchi N, Adachi R, Teraoka H. Toxicokinetics of a developmental toxicity test in zebrafish embryos and larvae: Relationship with drug exposure in humans and other mammals. Current Research in Toxicology. 2024;**7**:100187

[114] Ducharme NA, Reif DM, Gustafsson JA, Bondesson M. Comparison of toxicity values across zebrafish early life stages and mammalian studies: Implications for chemical testing. Reproductive Toxicology. 2015;**55**:3-10

[115] Miyawaki I. Application of zebrafish to safety evaluation in drug discovery. Journal of Toxicologic Pathology. 2020;**33**(4):197-210

[116] Shen C, Zuo Z. Zebrafish (*Danio rerio*) as an excellent vertebrate model for the development, reproductive, cardiovascular, and neural and ocular development toxicity study of hazardous chemicals. Environmental Science and Pollution Research. 2020;**27**(35):43599-43614

[117] Wang R, Liu K, Zhang Y, Chen X, Wang X. Evaluation of the developmental toxicity induced by E804 in zebrafish embryos. Frontiers in

Pharmacology. 2020;**11**:32. DOI: 10.3389/fphar.2020.00032

[118] Tal T, Yaghoobi B, Lein PJ. Translational toxicology in zebrafish. Current Opinion in Toxicology. 2020;**23-24**:56-66

[119] Zhang W, St Clair D, Butterfield A, Vore M. Loss of Mrp1 potentiates doxorubicin-induced cytotoxicity in neonatal mouse cardiomyocytes and cardiac fibroblasts. Toxicological Sciences. 2016;**151**(1):44-56

[120] Goodale BC, Du JKL, Bisson WH, Janszen DB, Waters KM, Tanguay RL. AHR2 mutant reveals functional diversity of aryl hydrocarbon receptors in zebrafish. PLoS ONE. 2012;7(1):e29346

[121] Xu Y, Peng T, Zhou Q, Zhu J, Liao G, Zou F, et al. Evaluation of the oxidative toxicity induced by lead, manganese, and cadmium using genetically modified nrf2a-mutant zebrafish. Comparative Biochemistry and Physiology Part C: Toxicology & Pharmacology. 2023;**266**:109550

[122] Souder JP, Gorelick DA. ahr2, but not ahr1a or ahr1b, is required for craniofacial and fin development and TCDD-dependent cardiotoxicity in zebrafish. Toxicological Sciences. 2019;**170**(1):25-44

[123] Zhang L, Kim SH, Park KH, Townsend DM, Tew KD, Zhi-wei Y, et al. Glutathione S-transferase P influences redox homeostasis and response to drugs that induce the unfolded protein response in zebrafish. The Journal of Pharmacology and Experimental Therapeutics. 2021;**377**(1):121-132

[124] Mech AM, Merteroglu M, Sealy IM, Teh MT, White RJ, Havelange W, et al. Behavioral and gene regulatory responses to developmental drug exposures in zebrafish. Frontiers in Psychiatry. 2022;**12**:795175

[125] Suzuki S, Kimura R, Maegawa S, Nakata M, Hagiwara M. Different effects of methylphenidate and atomoxetine on the behavior and brain transcriptome of zebrafish. Molecular Brain. 2020;**13**(1):70

[126] Demin KA, Lakstygal AM, Krotova NA, Masharsky A, Tagawa N, Chernysh MV, et al. Understanding complex dynamics of behavioral, neurochemical and transcriptomic changes induced by prolonged chronic unpredictable stress in zebrafish. Scientific Reports. 2020;**10**(1):19981

[127] Al-Soufi L, Arana ÁJ, Facal F, Flórez G, Vázquez FL, Arrojo M, et al. Identification of gene co-expression modules from zebrafish brain data: Applications in psychiatry illustrated through alcohol-related traits. Progress in Neuro-Psychopharmacology and Biological Psychiatry. 2024;**135**:111136

[128] Pavlidis M, Theodoridi A, Tsalafouta A. Neuroendocrine regulation of the stress response in adult zebrafish, *Danio rerio*. Progress in Neuro-Psychopharmacology and Biological Psychiatry. 2015;**60**:121-131

[129] Kundap UP, Kumari Y, Othman I, Shaikh MF. Zebrafish as a model for epilepsy-induced cognitive dysfunction: A pharmacological, biochemical and behavioral approach. Frontiers in Pharmacology. 2017;**8**:515. DOI: 10.3389/fphar.2017.00515

[130] Melgoza A, Guo S. Systematic screens in zebrafish shed light on cellular and molecular mechanisms of complex brain phenotypes. In: Molecular-Genetic and Statistical Techniques for Behavioral and Neural Research. Elsevier; 2018. pp. 385-400

Chapter 4

Mind the Gut: Overview on the Microbiota-Gut-Brain Axis Intricacies and Insights from the Zebrafish Model

Geonildo Rodrigo Disner, Carla Lima and Monica Lopes Ferreira

Abstract

The gut-brain axis represents a complex bidirectional communication network between the central nervous system and the gastrointestinal tract, notably through its microbiome. This axis involves multiple signaling pathways, including neural, hormonal, immunological, and metabolic processes, mediated by the intestinal microbiome that are highly conserved among vertebrates. These microorganisms play fundamental roles in modulating the immune system, digestion, and brain function through their interaction with the enteric nervous system, often referred to as the "second brain." Changes in the microbial community – known as dysbiosis – due to inflammatory diseases or extrinsic factors, such as diet and pollutants, can result in impairments in the development, physiology, behavior, and overall health of organisms, in addition to being identified as a contributing factor to the emergence of neurological and neurodegenerative diseases. Through studies using animal models such as zebrafish, new scientific advances can be achieved, driving the development of innovative strategies to prevent and treat human diseases, as well as a better understanding of the link between dysbiosis and neurological disorders. This approach addresses central issues of the United Nations Sustainable Development Goals, specifically those aimed at promoting health and well-being. Thus, this chapter aims to provide a comprehensive overview of the functions and implications of the microbiome-gut-brain axis in maintaining homeostasis, with a particular focus on the potential deleterious effects of dysbiosis and the insights gained in the last decades from using zebrafish as an experimental model.

Keywords: dysbiosis, microbiota-gut-brain Axis (MGBA), microbiome, neurobehavior, zebrafish

1. Introduction

Humans and other animals are remarkably complex systems whose functioning, well-being, and homeostasis are shaped by a delicate interplay of intrinsic and extrinsic factors. Among the intrinsic factors, genetics, metabolism, and the immune system play critical roles, while extrinsic influences, such as environment, diet, and lifestyle, further modulate most biological processes. Collectively, these elements converge to shape an individual's health and overall well-being.

In recent years, microbiome research has unveiled how microbial communities (including bacteria, fungi, and viruses residing in and on the human body) serve as key mediators between built-in and outside components, influencing everything from immunity to metabolism. Among these microbial ecosystems, the gut microbiota (i.e., the bacteria inhabiting the digestive system) has emerged as a mighty player, exerting profound effects on human health. These microorganisms play essential roles in immune modulation, digestion, and even brain function, underscoring their far-reaching impact [1]. This rapidly evolving field highlights the profound interconnectedness between human biology and the microbial world, offering transformative insights for health, disease, and personalized medicine.

Contemporary research evidence suggests a strong biological connection between the mind and the digestive system [2, 3]. Studies at the forefront of knowledge have demonstrated an intricate connection between dysbiosis – an imbalance in the composition of the microbiome – and the emergence of psycho-neurological diseases, such as depression, autism spectrum disorder (ASD), and neurodegenerative diseases [4, 5]. Thus, this connection between the microbiome and the nervous system, mediated by the microbiome-gut-brain (MGB) axis, represents a promising area that has been increasingly explored. In addition, it enables a greater understanding of how exposure to chemical compounds or habits, such as a healthy diet and proper hydration, can influence physical health and psychological well-being.

Through studies using animal models such as zebrafish (*Danio rerio*), which allow the recapitulation of changes in the microbiome-gut-brain axis, new scientific advances can be achieved. Such findings include the development of treatments based on the modulation of the microbiome or screening for compounds that have an adverse effect on the intestinal microbial community, which can induce neurological and inflammatory diseases, paving the way for innovative strategies for the prevention and treatment of human diseases [4, 6, 7].

Thus, research that seeks to elucidate the relationship between disruption of the microbiome-gut-brain axis and the appearance of atypical physiological and behavioral manifestations has the potential to revolutionize areas such as neuroscience, immunology, psychiatry, and even personalized medicine, contributing to a more integrative understanding of human health, as well as potentially teaching us that, with minor changes in our diet or lifestyle, we can enjoy a calmer mental state, improve immunity, and reduce the risk of developing neurological diseases such as Parkinson's and Alzheimer's. Ultimately, the balance between the mind and gut may be the key to achieving more vigorous health.

Therefore, this chapter aims to provide a comprehensive overview of the functions and implications of the microbiome-gut-brain axis in maintaining homeostasis, with a particular focus on the systems-level relationships, potential deleterious effects of dysbiosis, and the insights gained from using zebrafish as an experimental model in this field of research.

2. The microbiota-gut-brain (MGB) axis

The gut-brain axis has long been known as essential for maintaining homeostasis. The intricate and complex interaction between the gut and the brain has become an increasingly important topic in contemporary science due to its profound health implications [8]. Nonetheless, the gut microbiota, a vast community of trillions of microorganisms, has been recognized over the past two decades as an essential component in the regulation of the gut-brain functions. This progress has underscored the importance of a distinct MGB axis, a framework now widely acknowledged in the scientific literature. Mounting evidence has unveiled the bidirectional communication between the gut microbiome and the central nervous system (CNS), further solidifying the critical influence of this axis on physiology and behavior [2, 3, 8].

The MGB axis operates through bidirectional communication. This crosstalk occurs via multiple pathways, including immune signaling, vagus nerve transmission, and the enteric nervous system, as well as through microbial metabolic activity, such as the production of short-chain fatty acids, branched-chain amino acids, and peptidoglycans. Together, these mechanisms integrate gut microbial signals with CNS processes, underscoring the pivotal role of this axis in host physiology [9].

As reviewed by Cryan *et al.* [9], multiple lines of evidence support the idea that the microbiota plays a significant role in modulating the gut-brain axis. First, studies using germ-free animals have demonstrated that the absence of gut microbes results in noticeable alterations in brain development and function. Second, behavioral changes observed in animals administered specific bacterial strains, and the confirmation of similar effects in human studies, highlight the potential for cross-species translatability. Third, population-based studies, such as those following the Walkerton outbreak in Canada (*Escherichia coli* contamination), have demonstrated that exposure to infections can lead to lasting gut-brain symptoms. These findings are supported by animal models, where even low-level infections can influence behavior without triggering an immune response. Fourth, preclinical research has demonstrated that administering antibiotics, whether early in life or during adulthood, can lead to long-term alterations in the brain, spinal cord, and enteric nervous system. Finally, clinical observations of hepatic encephalopathy reveal that targeting the gut microbiota with antibiotics can significantly improve neurological symptoms. Collectively, this understanding shaped and strengthened the concept of the MGB axis, which is now a central focus in neuroscience and microbiome research.

Due to its critical role, the microbiome-gut-brain axis emerges as a promising platform for developing novel therapeutics interventions. Such treatments could address a wide spectrum of disorders, from conditions affecting mental health and cognitive function to obesity and various gastrointestinal (GI) illnesses, notably inflammatory bowel disease (IBD) and irritable bowel syndrome (IBS).

Zebrafish possess a functionally conserved MGB axis that shares fundamental mechanisms with humans, while offering unique experimental advantages [10]. As in humans, the zebrafish MGB axis operates through bidirectional communication pathways involving microbial metabolites, immune signaling, and neural connections. These shared pathways similarly influence neurodevelopment, stress responses, and behavior in both species. The zebrafish gut microbiota, like its human counterpart, produces neuroactive compounds that can modulate CNS function, demonstrating remarkable evolutionary conservation of these interactions. Importantly, key microbial metabolites, such as butyrate and serotonin precursors, exhibit comparable neuroactive effects across both species. However, zebrafish exhibit faster microbiota colonization and simpler neural circuitry, allowing more precise observation of direct microbiota-brain effects without the complexity of mammalian cortical modulation [10, 11]. Despite these differences, the high degree of functional conservation makes the zebrafish a powerful model for studying fundamental mechanisms of the MGB axis, which remain relevant to human biology, particularly in early neurodevelopment and stress-related disorders. The species' genetic tractability and transparency further enhance its utility for dissecting causal relationships in gut-brain communication that are difficult to study in humans.

Overall, the zebrafish is a highly valuable model for translational research. Both larval and adult zebrafish offer unique insights into disease mechanisms and drug responses, while their sensitivity to neuroactive compounds mirrors human pharmacodynamics in key aspects. The expanding molecular toolkit available for zebrafish manipulation continues to refine its applicability in modeling pathologies with increasing precision. As the use of the zebrafish model continues to expand, its contribution to identifying therapeutic targets, validating biomarkers, and guiding the development of safe, high-efficacy drugs will become increasingly indispensable, securing its role as a cornerstone in translational science [12–14].

3. Microbiome and the enteric nervous system

Our planet is mainly a microbial world. The number of microbial species on the Earth vastly exceeds that of all other life forms combined. Thus, our existence is profoundly shaped by microbes, which are ancient, invisible life forms that have thrived for hundreds of millions of years, long before humans emerged. Therefore, there has not been a moment in our history when our bodies have not been in constant dialog with these microscopic partners, receiving their signals and responding in ways that continue to define our biology, health, and even our minds.

The microbiota is defined as the set of microorganisms that inhabit the body. In the context of human health, the microbiota is often associated with the colon microbiota, which harbors the majority (~95%) of the body's bacteria. In contrast, the microbiome encompasses both microorganisms and their genes, as well as the biological interactions they facilitate. Collectively referred to as the "second human genome," the intestinal microbiome is considered by some experts to be a separate "organ" with distinct metabolic and immunological activity [15].

Interestingly, there are about 10 times more microbial cells in the gut than in the entire human body, comprising approximately 100 trillion microscopic organisms, within up to 5000 different species, with an approximate mass of about 1 kilogram [8, 15, 16]. The estimates, while potentially overestimated, are impressive. Some divergences have emerged in recent years, such as the work of Sender *et al.* [17], which revisited the question of the ratio of bacterial to host cells in humans, finding—based on more accurate and updated quantitative criteria—that it is much closer to 1:1, despite often presented as common knowledge that bacteria outnumber human cells by a ratio of at least 10:1.

Most of them are harmless or beneficial to the host, acting as protection against enteropathogens, extracting nutrients and energy from the diet, and contributing to immune functions. Therefore, understanding how the gut microbiota affects health and disease requires a shift in focus from individual pathogens to an ecological approach that considers the community as a whole [8].

Viewed through this lens, it has become clear that the microbiota plays a central role in shaping human health and disease, serving as a central regulator of host physiology. To integrate this idea, the concept of the holobiont has been developed to describe the ecological unit comprising both the host species and its symbiotic microbiota [18].

Although each person's microbiota profile is distinct, studies suggest that all healthy adults share a group of intestinal bacterial species, which would constitute a "core microbiota." It is estimated that each person harbors thousands of phylotypes at the species level, with the two most prominent phyla being Firmicutes and Bacteroides, representing at least ¾ of the microbiome, while the most prevalent species are *Faecalibacterium prausnitzii*, *Roseburia intestinalis*, and *Bacteroides uniformis* [8, 19]. According to combined data from the Human Microbiome Project and MetaHIT, at least 2776 prokaryotic species have been identified in human fecal samples. These species span 11 distinct phyla, with *Proteobacteria*, *Firmicutes*, *Actinobacteria*, and *Bacteroidetes* collectively representing over 90% of the gut microbiome [20, 21]. In contrast, *Fusobacteria* and *Verrucomicrobia* are present in much lower abundances [22].

A vast and dynamic microbial diversity has been observed, varying significantly over time and across different human populations [8]. Several factors can influence the microbiota composition during an individual's life, including infections, mode of birth delivery, use of antibiotics, nutritional provision, environmental stressors, and host genetics. Additionally, microbial diversity tends to diminish with aging [9].

In zebrafish, on the other hand, studies have shown that, under controlled laboratory conditions, the diversity and abundance of the most common microbial phyla tend to be relatively uniform, reflecting a stable core microbiota. However, this apparent uniformity can be significantly influenced by a range of variables, including the origin of the animals, their genetic background or lineage, dietary composition, and various environmental factors such as housing systems, water sources, and microbial exposure. According to Borrelli *et al.* [23], an analysis of gut bacterial communities reveals that they are dominated by *Fusobacteria*, followed by *Proteobacteria*, *Firmicutes*, *Actinobacteria*, and *Cyanobacteria*. This diversity is also evident in the literature, as illustrated by Davis *et al.* [11], who found that, at 6 days post-fertilization (dpf) in zebrafish larvae, the phyla *Proteobacteria* and *Cyanobacteria* were the most prevalent. Teng *et al.* [24] found *Proteobacteria*, *Bacteroidetes*, *Firmicutes*, and *Fusobacteria* as the most abundant in the juvenile zebrafish intestine, while Shi *et al.* [25] reported the *Firmicutes*, *Proteobacteria*, *Fusobacteriota*, and *Bacteroidota* as most abundant phyla in zebrafish guts under non-treated conditions.

Since the gut is the microenvironment where most of the interactions between the microbiota, their metabolites, and the host happen, it soon got clear to the scientists that such interaction prevalently happens through a network of neurons known as the enteric nervous system (ENS), positioned to respond directly or indirectly to the microbiota and its metabolites [9]. This interconnection, known as the MGB axis, represents a tangled physiological network of bidirectional communication involving neuroendocrine, immunological, and neural pathways. Notably, the gut serves as a crucial interface where microorganisms, their metabolites, and the immune and nervous systems interact [26].

This "small brain," specifically, is a complex network of more than 100 million nerve cells that extends through the gastrointestinal tract, popularly known as the "brain in the gut" or "second brain" due to its autonomy and vast communication with the CNS [27]. The ENS is structured into two ganglionated plexi, the submucosal and myenteric plexus, containing neurons and enteric glial cells, connected by a dense network of nerve fibers and glial processes [28]. It is primarily responsible for regulating intestinal motility, the secretion of digestive enzymes, immune function, the growth and integrity of the mucosa, intestinal permeability, blood flow in the intestine, and interacting directly with the microbiome [28].

ENS is currently recognized as playing a crucial role in general health and gastrointestinal physiology, affecting inflammatory and behavioral processes, as well as being related to diseases such as irritable bowel syndrome and neuropsychiatric disorders, through the microbiome-gut-brain axis [1, 19, 27, 29–31]. Several recent studies have demonstrated that the microbiota also plays an important role in the development of brain function, pathology of neurobehavioral, psychiatric, and neurodevelopmental disorders, some of which depend on activation of the vagus nerve [16, 32–36].

Cryan *et al.* [9], Hyland and Cryan [37], and other groups have reviewed the mechanisms by which the microbiota influences ENS development and function, which includes activation of pattern recognition receptors (PRRs), including toll-like receptors (TLRs), in particular, TLR2 and TLR4, which are involved in the recognition of microbial molecules. While much attention has been given to how the microbiota influences ENS development and function, it is now becoming clear that the ENS can, in turn, regulate the composition and activity of the gut microbiota [38]. Additionally, studies in germ-free animals have shown that bacterial colonization of the gut is critical for the development and maturation of the enteric and central nervous systems [30, 39].

The brain and gastrointestinal tract are both essential sensory systems that identify, relay, process, and react to cues from the body's internal and external surroundings. At the interface of this sensory function, immune cells in the gut and brain consistently track environmental factors, eliciting responses that inform the body's physiological state [40]. Thus, understanding aspects related to the microbiome-gut-brain axis provides valuable insights into its roles in maintaining homeostasis, overall health, mental well-being, and the sustainable balance of the intestinal microbiota.

4. Dysbiosis and its downstream effects

Dysbiosis is broadly understood as the disruption of the normal balance between the microbiota and the host, marked by the change in the community diversity—usually characterized by the loss of beneficial microorganisms and expansion of potentially harmful ones—or abundance shift of specific microbial species [32, 35, 41]. Parallel and clinically oriented definitions can also be represented by compositional and functional alterations in the microbiota in individuals with disease compared to healthy subjects [42]. However, it should be noted that, more often than not, unhealthy outcomes evolve as a consequence of dysbiosis rather than being directly caused by it.

One of the most valuable benefits conferred by a balanced microbiota is the ability to repulse pathogens, which is known as colonization resistance [43]. This phenomenon is part of the intrinsic roles the microbiota play in supporting the balance and well-functioning of the whole system. While facing minor microbiota changes, individuals may begin to exhibit early symptoms of related diseases, such as immunosuppression, an inflammatory response, or even changes in behavior and well-being.

During inflammation triggered by or resulting from dysbiosis, immune and non-immune cells form complex networks of interaction to regulate such responses in the central nervous system, enteric system, and gastrointestinal tract [40]. Many neurological disorders are associated with gut inflammation and inflammatory markers in the blood, which are linked to deficits in brain function and behavior. Specifically, inflammatory bowel diseases are a group of inflammatory conditions unrelated to infection, and their prevalence is increasing significantly in industrialized countries. Advances in the understanding of these comorbidities and their relationship with dysbiosis have been made through chemical induction models of the disease. Some of the compounds most frequently used to induce dysbiosis in animals are sodium dextran sulfate (DSS) and 2,4,6-trinitrobenzene sulfonic acid (TNBS) [44]. Although the exact mechanism of DSS-induced colitis remains unclear, it is thought to have a deleterious effect on intestinal epithelial cells, leading to the disruption of barrier integrity and the translocation of commensal bacteria, which then activate innate immune responses [7, 44].

In addition to chemicals, other environmental factors can also exert a negative effect on the intestinal microbial profile. In this sense, diet appears to have one of the most potent influences on the intestinal microbiome. In modulating dietary dysbiosis, prolonged exposure to a high-fat diet induces a significant increase in the expression of markers of pro-inflammatory, apoptotic, and proliferative processes in the tissues of the intestine, liver, and brain [45]. Long-term consumption of a high-fat diet can even cause activation of the Wnt/β-catenin signaling pathway in brain tissue. Fish fed with such a diet were shown to present anxiety, aggression, and hyperactivity [46], which confirms that an imbalance in intestinal homeostasis and the microbiome can lead to non-traumatic brain injury and stimulate a regenerative response.

In the literature, considerable attention is devoted to investigations focused on understanding the association between the microbiota and a wide range of human pathologies, including metabolic, inflammatory, neurological diseases, and even cancer [41]. Despite this growing attention, further studies are needed to deepen our understanding of the specific mechanisms linking the microbiota to various pathologies and to develop effective therapeutic solutions. These investigations should focus on how the microbiota directly contributes to the development or exacerbation of these conditions.

Lastly, the assessment of how dysbiosis can be correlated to health impairments was driven by advancements in sequencing techniques. Usually, the dysbiosis is evaluated by sequencing the V3 and V4 regions of the 16S rRNA gene. This is the most commonly used target, as it serves as a universal marker for bacterial identification and classification. The gene contains highly conserved regions and hypervariable regions, providing quantitative information to classify microbial communities.

5. Zebrafish as a model for microbiome-gut-brain axis research

Animal models have been fundamental in the study of the cellular and molecular mechanisms underlying human diseases, providing the basis for understanding animal physiology and developing new therapeutic strategies [44]. Specifically, regarding the MGB axis, the use of relevant laboratory animals with varying levels of microbial complexity presents a promising opportunity to better understand the effects of host-microbiota interactions in regulating health and disease states. Furthermore, manipulation of the composition of the gut microbiota, the use of probiotic treatments, and even the creation of germ-free strains in vertebrates and invertebrates allow for functional and mechanistic studies aimed at determining causality in disease-associated biological changes [41].

In this context, zebrafish have emerged in recent decades as a valuable laboratory animal for studying the MGB axis, with significant implications in the areas of translational medicine, toxicology, disease modeling, and behavioral studies. This fish, which serves as an alternative model, offers numerous advantages that make it an ideal candidate for unraveling the complex interactions between the gut microbiome, the nervous, and immune systems [47–50]. Furthermore, since the first applications by George Streisinger in the early 1980s, zebrafish have become a prominent organism for understanding the role of leukocytes in controlling sterile or microbial-induced inflammation, for modeling various human diseases, as well as in pharmacology through studies that contribute to toxicological assessment and preclinical drug discovery [51–53].

The zebrafish is a small freshwater fish native to South Asia (3–4 cm in length), which is generally considered a juvenile at 45 dpf and an adult at 90 dpf. Females spawn every 5–7 days, laying an average of 300 eggs per week, and reproduce year-round. The eggs are fertilized ex vivo, and the transparent embryos and larvae develop ideally at an average temperature of 28°C [54].

Zebrafish embryogenesis is rapid with a distinct head-tail axis evident by 10 hours post-fertilization (hpf), and the process is characterized by the period of cleavage (45–90 min), blastula (4 h), gastrulation (10 h), segmentation (10.5–20 h), pharyngeal period (24–36 h), and hatching (48–96 h) [55]. Most major organs, including blood vessels, brain, heart, liver, intestine, and eyes, are formed by two dpf, becoming fully functional within a week [56]. The process of organ formation, such as the kidneys, heart, blood-brain barriers, and gastrointestinal system, also occurs similarly in mammals [57].

Specifically regarding the zebrafish gastrointestinal tract, its intestine forms a folded tube that fills most of the abdominal cavity. While it possesses only three intestinal segments, unlike the five found in mammals, there is significant functional similarity [58, 59]. Instead of a stomach, teleost fish have an anterior segment, called the intestinal bulb, characterized by an expanded lumen and strong expression of digestive enzymes [58]. The intestinal bulb is followed by the midgut, which contains goblet and enteroendocrine cells, and then a short posterior section, which is devoid of absorptive enterocytes, analogous to the colon in mammals [59].

The intestinal epithelium of zebrafish exhibits similarities to that of mammals (**Figure 1**). Intestinal epithelial cells travel from the base of the folds up to the tip, culminating in their elimination through apoptosis. As in mammals, the renewal of these epithelial cells depends on the microbiota [61]. There are also similarities in the mechanisms that control differentiation toward the secretory lineage, such as the conservation of Delta-Notch signaling to induce goblet cell differentiation and the conserved inhibition of Kruppel-like factor 4 (KLF4), a master regulator of goblet cell differentiation, by retinoic acid [62, 63]. Immune cells can also be detected within the intestine, where T cells are present from 3 weeks post-fertilization [64]. Zebrafish also exhibit several differences in their intestinal epithelial architecture compared to mammals. For example, the epithelium lacks Lieberkühn crypts, and only three types of differentiated epithelial cells have been identified in zebrafish: enterocytes, goblet cells, and enteroendocrine cells [65].

Microbial colonization of the zebrafish begins around three dpf, when larvae hatch from the axenic environment within the chorion [59]. By 2 weeks of age, the zebrafish microbiome is relatively well established, with a community that resembles that found in adults. The core microbiota undergoes specific changes over time, being dominated by *Proteobacteria* (76–82%), a notable difference from mammals, where *Firmicutes* and *Bacteroidetes* are the dominant phyla. Qualitatively, six of the eleven bacterial phyla found in zebrafish are detected in the mouse gut microbiota, and five

DOI: http://dx.doi.org/10.5772/intechopen.1014305

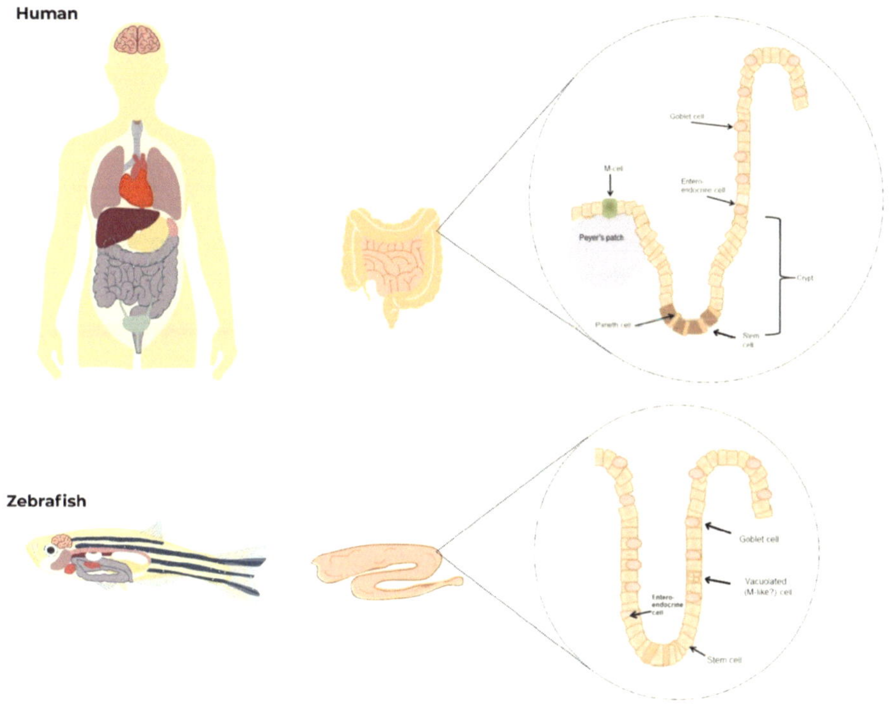

Figure 1.
Compared anatomy of the human and zebrafish intestinal tracts and its epithelial structure. Zebrafish do not have Paneth cells, crypts, and organized lymphoid structures such as mesenteric lymph nodes, isolated lymphoid follicles or Peyer's patches. Comparison of cell types and intestinal structures (detail) was adapted from Brugman [60].

of these are also shared by the adult human microbiota [66]. On the other hand, the microbiota of zebrafish and mammals respond similarly to various stimuli, including toxins or high dietary levels of fat, which allows for a significant understanding of the subject from studies using the zebrafish model [41].

Regarding general aspects of the immune response, zebrafish exhibit broad conservation in the response of cytokines, receptors, and components of several signaling pathways, primarily those known for recruiting cells of the innate immune system, such as neutrophils and macrophages. A phylogenetic study of cytosolic NOD-like receptors (NLRs) in zebrafish revealed three distinct families: NLR-A, orthologs of mammalian NOD-like receptors; NLR-B, orthologs of mammalian NLRP receptors; and NLR-C, a group present exclusively in teleost fish [67, 68].

Given the in-depth knowledge of zebrafish biology and physiology, as well as their high homology with humans, it is possible to establish models of both inflammatory bowel and neurological diseases. In the context of inflammatory bowel diseases, zebrafish have emerged as a functional model to study the inflammatory response through disruption of the MGB axis, and also allow us to explore how this altered signaling network contributes to metabolic and neurological disorders. In the latter case, research with zebrafish allows researchers to study neurodegenerative diseases – such as Parkinson's and Alzheimer's – as well as biological responses, including anxiety, stress, or hyperactivity, as they exhibit behavior comparable to that of mammals, but are easier to handle and reproduce through behavioral tests, such as the exploration test [23, 69].

The research focused on the zebrafish MGB axis, evaluation of dysbiosis, and its consequences have been extensively studied using zebrafish in recent years. An overall summary of contemporary research in the main areas is presented below.

5.1 Experimentally induced shifts in the microbiota

The study of experimentally induced dysbiosis has been useful to elucidate the effects and mechanisms related to the alteration in microbial diversity associated with chemical exposure or rearing settings. In that regard, dysbiosis can be tested primarily by drug administration, different diet regimes, or exposure to environmental contaminants.

It has been well established in the literature that overnutrition and calorie-dense foods have the potential to alter lipid metabolism, induce inflammation, and alter gut microbiota. The approach, consisting of providing high-fat diets to zebrafish and, after a medium to long period, assessing the effects on dysbiosis and inflammatory markers, is now often applied in experimental research. Some examples to illustrate this methodology are presented here. For example, Arias-Jayo *et al.* [45] established a high-fat diet in a zebrafish colony over a period of 25 days. The exposure led to microbial dysbiosis, marked by an increased relative abundance of the phylum *Bacteroidetes*, with additional activation of NF-κB signaling, which confirmed the onset of intestinal inflammation, also evidenced by the upregulation of genes involved in receptor signaling, antimicrobial responses, and the inflammatory cascade. The authors find that the intestinal barrier was compromised, accompanied by elevated mucin production from goblet cells.

In addition, Qiao *et al.* [70] also reported the decreased abundance of *Proteobacteria* in the zebrafish fed with a high-fat diet (15% fat compared to 7% fat in the control group). Interestingly, when authors further investigate the causal relationship between the microbiota and lipid accumulation in colonized three dpf germ-free zebrafish and the intestinal microbiota of adult zebrafish from the high-fat diet experiment, they identify that such microbiota reshaping leads to detrimental lipid accumulation.

Recently, zebrafish have been widely used to investigate potential agents of dysbiosis, particularly in the form of pollutants, besides drugs and diets. This area connecting environmental toxicology with dysbiosis and physiological impairments in zebrafish has been very fruitful. Examples of studies that illustrate such line of research are the investigation of pollutants such as (nano)microplastics, where significant alterations in the gut microbiome and tissue metabolic profiles were observed, with most of these alterations linked to inflammation, lipid metabolism, and oxidative stress [31, 71]. Also, nanomaterials, like quantum dots, induced behavioral abnormalities in zebrafish, associated with inhibited acetylcholinesterase activity, increased dopaminase activity, and neuronal apoptosis. In the intestine, the exposure resulted in disrupted mucosal barriers, increased permeability, and triggered oxidative stress and apoptosis, as well as a reduction in gut microbial diversity [72]. Finally, literature findings indicate that exposure to per- and polyfluoroalkyl substances (PFAS) can induce hepatic immunotoxicity, associated with intestinal microbiota dysbiosis, in adult zebrafish [73].

Due to the elevated use of psychoactive therapeutic drugs, their release into the environment has also been of crucial concern. Flunitrazepam, for example, is a medication that causes severe depression of the CNS and is used sporadically to reduce anxiety attacks and sleep disorders. In that regard, Lin *et al.* [48] investigated how a 30-day exposure to the benzodiazepine drug flunitrazepam (FLZ, 0.2 and 5 μg/L) induces neurotoxicity. The authors found that FLZ disrupted the balance of gut microbiota and caused an increase in pathogenic bacteria, such as Paracoccus and Aeromonas, leading to pathological damage to the intestine. Moreover, intestinal pro-inflammatory factors (i.e., IL-1β and TNF-α) were upregulated, while tight junction proteins, Occludin and Zonula Occludens 1, were found to be downregulated.

Wang *et al.* [49] also studied the mechanisms of toxicity of triclosan, a broad-spectrum bactericide highly used in the industry, on the gut-brain axis. For that purpose, zebrafish were exposed to 250 μg/L of triclosan from 30 to 90 dpf. The chronic exposure to triclosan affected the zebrafish development and caused neurobehavioral abnormalities by inhibiting acetylcholinesterase and dopamine activity, promoting brain neuron apoptosis, and accelerating diencephalic lesions. Moreover, the exposure led to gut microbiota dysbiosis and decreased diversity, characterized by an increase in pathogenic bacteria and a decrease in probiotics in the adult zebrafish gut, resulting in numerous pathological damages, including inflammatory infiltration, thinning of the intestinal wall, and an increase in goblet cells within the villus.

Novel classes of environmental pollutants, such as the (nano)microplastics, have been implicated in zebrafish dysbiosis. To investigate the adverse effects of 50 μm and 100 nm (nano)microplastics on the zebrafish gut microbiota, antioxidant activity, and innate immune response, Pei *et al.* [71] exposed adult zebrafish to 100 and 1000 μg/mL of the xenobiotic. The exposure led to significant changes in the richness and diversity of gut microbiota, inducing oxidative damage through the increased activity biomarkers such as superoxide dismutase (SOD), catalase (CAT), and glutathione peroxidase (GSH-Px). It altered the level of most pro- and anti-inflammatory factors, including IL-8, NF-κB, IL-10 (increased), and TNF-α (decreased). Still, in that regard (i.e., nanotoxicology), Cao *et al.* [72] tested if and how quantum dots (Black Phosphorus Quantum Dots - BPQDs, 5, 50, and 500 μg/mL) could induce toxicity and dysbiosis in zebrafish. The authors observed that, following 35 days of exposure, the substance caused locomotor behavior abnormalities, inhibited acetylcholinesterase activity, induced dopaminase activity, and promoted apoptosis in the brain tissue. Meanwhile, BPQDs caused damage to the physical and chemical barriers in zebrafish intestinal tissue, which increased the permeability of the intestinal mucosa, and induced oxidative stress and apoptosis. Additionally, dysbiosis was observed as a result of the decreased diversity. Specifically, the relative abundance of *Firmicutes*, *Bacteroidetes*, and *Actinobacteria* decreased, while the relative abundance of *Proteobacteria* and *Clostriobacteria* increased.

Another example of how contaminants of emerging concern (CEC) could impact the MGB axis is illustrated by the study of Huang *et al.* [73] when assessing the perfluorooctane sulfonate (PFOS) toxicity for 21 days. In summary, the results support the role of PFOS in changing the relative abundance of *Plesiomonas*, *Aeromonas*, *Cetobacterium*, *Shewanella*, and *Vibrio*. Furthermore, PFAS exposure upregulated the expression of *hepcidin*, *muc*, and the immune-related genes *mpo* and *saa*, while downregulating the tight junction-associated gene *occ* in the intestine.

5.2 Use of probiotic treatments

Some bacteria are termed psychobiotics because of their ability to alleviate stress and anxiety-related behaviors, thereby enhancing mood and cognition [4]. This approach, which involves introducing specific strains to promote well-being and restore homeostasis, falls within the broader field of probiotics.

To test the effects of probiotic treatments, Olorocisimo *et al.* [4] tested three Lactobacillaceae isolates, namely, *Lactobacillus paracasei*, *Lactobacillus delbrueckii*, and *Lactobacillus casei*, which were inoculated twice a day into fish tanks containing 10 6–7 month-old female fish each, at a final concentration of 1.6×10^6 CFU/mL per aquarium for 2 weeks. Authors reported a considerable decrease in anxiety-like behavior in *L. delbrueckii*-exposed fish, as assessed by the novel object exploration and novel tank diving tests. This observation resulted from the fact that the *L. delbrueckii* group spent most of the time in the upper half of the novel tank and had the shortest latency time. Besides the behavioral tests, the expression patterns of genes commonly associated with anxiety reveal that glutamic acid decarboxylase (*gad1*) was upregulated in the gut of *L. delbrueckii* group, as well as in the brain of *L. casei*-exposed group. To confirm that such an effect is due to a change in bacterial taxa abundance, microbial sequencing revealed that the *L. delbrueckii*-treated group showed a lower percentage of *Firmicutes* and a higher percentage of *Actinobacteria* than controls. Genus-level analysis revealed that zebrafish exposed to *L. delbrueckii* exhibited a marked increase in the abundance of *Lactobacillus* sp., *Verrucomicrobium* sp., *Kaistia* sp., *Methylobacterium* sp., *Singulispira* sp., *Gordonia* sp., and *Mycobacterium* sp. Conversely, the genera *Legionella*, *Planctomyces*, *Flavobacterium*, and *Prevotella* showed a reduction in abundance compared to controls.

Another study focused on investigating the probiotic modulation outcomes in zebrafish behavior was the Borrelli's *et al.* [23] study. After 28 days of dietary administration with the probiotic *Lactobacillus rhamnosus* IMC 501 it was found differences in shoaling behavior, brain expression levels of *bdnf* and of genes involved in serotonin signaling and metabolism (i.e., *tph1a*, *tph1b*, and *htr1aa*) between control and treated zebrafish group. In addition, there was a reported modulation in the microbial communities with a significant increase of *Firmicutes* and a trending reduction of *Proteobacteria*.

5.3 Axenic (germ-free) and gnotobiotic strains

Zebrafish reproduction is well known, and embryo development occurs outside the progenitor body, which provides crucial advantages to early life studies. More than that, the interior of the chorion is free of microorganisms, providing an excellent opportunity to use this species to answer scientific questions where a germ-free status may be required. Given the comparative simplicity of producing germ-free zebrafish, this species represents a compelling model for studies focused on gnotobiotics.

In the scientific context, axenic status refers to a condition in which an organism or culture is established and sustained in the strict absence of any other contaminating organisms, that is, it refers to a state of being free from any association or contamination with other organisms. In comparison to the axenic strain, there are the gnotobiotic colonies (from Greek *gnōtos* 'known' + biotic). This term describes animals in which the microbial status is fully known. Within this broad definition, one can consider two main subclasses: (I) mono-associated, that is, animals colonized with only one defined microbial species; and (II) poly-associated, that is, colonized with a defined (but reduced) community of microbes. It is important to note that these two approaches are correlated, as the starting point for all gnotobiotic zebrafish research is the derivation of germ-free embryos (**Figure 2**) [74, 75].

Since zebrafish behavior can be easily tracked in the laboratory, scientists can now associate alterations in these patterns with microbial changes, linking these effects to neurodegenerative disorders' phenotypes. One study led by Weitekamp and

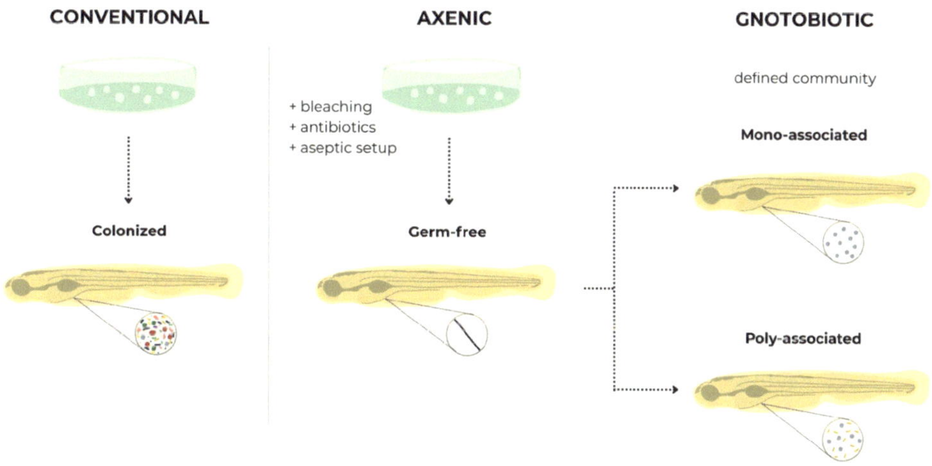

Figure 2.
Colonization status of zebrafish. The chorion acts as a protective barrier that is largely impermeable to microbes in the surrounding water. This characteristic makes zebrafish embryos an excellent model for studying host-microbiome interactions, as researchers can easily generate and maintain germ-free (axenic) fish and colonies with defined communities of microorganisms (gnotobiotic).

Kvasnicka [76] demonstrates a clear relationship between zebrafish dysbiosis and behavioral alterations, specifically hyperactivity. Axenic zebrafish larvae exhibited increased locomotor activity. This phenotype was reversed when associated with particular host-derived bacterial species or their mixtures, suggesting that specific microbes are essential for normal neurobehavioral development. Moreover, the findings highlight that the composition and complexity of the intestinal microbiota play a critical role in modulating zebrafish behavior, with implications for experimental reproducibility in behavioral studies involving pharmacology and toxicology.

In another study, scientists tested whether the microbial composition could influence the zebrafish's developmental neurobehavioral toxicity due to exposure to the hazardous chemical benzo[a]pyrene (B[a]P). Stagaman *et al.* [6] applied a gnotobiotic zebrafish experiment, up to 9 dpf, and found that gut microbiome state altered the relationship between B[a]P exposure concentration and behavioral response.

Additionally, indirect effects resulting from the disruption of beneficial host-microbiota interactions may also contribute to the toxicity of nanoparticles, notably silver nanoparticles, which have antimicrobial capacity. In that context, Brinkmann *et al.* [77] compared the acute toxicity of silver and zinc oxide nanoparticles (nAg and nZnO) to zebrafish larvae that were either germ-free or colonized by microbiota. Over 2 days of exposure, germ-free zebrafish larvae were more sensitive to nAg than microbially colonized larvae, whereas silver ion toxicity did not differ between germ-free and colonized larvae.

6. Final considerations and perspectives

In a world where novel chemicals are continuously being produced and the exposome (i.e., the totality of environmental exposures over a lifetime) is continually expanding, microbiome research and its effects on the gut-brain axis highlight the urgent need to understand how emerging contaminants and lifestyle habits impact host physiology through microbiome disruption.

Furthermore, although significant progress has been made in terms of scientific knowledge, we still have a long journey to go to understand the full, body-wide impact of a shifting microbiome. Because we have co-evolved with our microbiota, they play a crucial part in setting up and running all our other bodily systems, meaning that microbial communities are not passive guests but active regulators of host physiology, particularly influencing immune maturation, neurodevelopment via the gut-brain axis, and metabolic processes. Disruption of this microbial ecosystem, through factors such as antibiotics, poor diet, or environmental contaminants, can lead to systemic dysregulation and contribute to a range of disorders. This understanding positions the microbiota as a system-wide modulator of health and disease, emphasizing its relevance in avant-garde scientific research seeking to find a deeper understanding of disease progression and homeostasis.

Building on this perspective, it is important to recognize that while there is growing evidence linking microbiota alterations to psychiatric, neurological, and neurodegenerative diseases, definitive causal relationships remain unproven and require cautious interpretation. As toxicology shifts toward a systems-level, mechanism-driven approach, integrating microbiome research in biomedical research will be essential to unravel how exposures shape host health through microbial pathways. Continued multidisciplinary efforts are crucial for moving beyond speculation and translating microbiome insights into practical strategies for disease prevention and health promotion.

Acknowledgements

GRD was supported, in part, by the Brazilian National Council for Scientific and Technological Development – CNPq – through the program "Conhecimento Brasil" (process number #446270/2024-5 and #316193/2025-9). MFL and CL thank the Butantan Foundation for the support. The authors acknowledge the use of Grammarly for the language polishing of the manuscript.

Authors contribution

All authors have made substantial intellectual contributions and approved it for publication.

Conflict of interest

The authors declare no conflicts of interest.

Author details

Geonildo Rodrigo Disner[1,2]*, Carla Lima[1] and Monica Lopes Ferreira[1]

1 Plataforma Zebrafish of the Laboratory of Applied Toxinology, Butantan Institute, São Paulo, Brazil

2 Pesquisador Conhecimento Brasil do CNPq, São Paulo, Brazil

*Address all correspondence to: disner.rodrigo@gmail.com

References

[1] Thursby E, Juge N. Introduction to the human gut microbiota. The Biochemical Journal. 2017;**474**(11):1823-1836. DOI: 10.1042/BCJ20160510

[2] Schneider KM, Blank N, Alvarez Y, Thum K, Lundgren P, Litichevskiy L, et al. The enteric nervous system relays psychological stress to intestinal inflammation. Cell. 2023;**186**(13):2823-2838.e20. DOI: 10.1016/j.cell.2023.05.001. Epub 2023 May 25

[3] Wang Q, Yang Q, Liu X. The microbiota-gut-brain axis and neurodevelopmental disorders. Protein & Cell. 2023;**14**(10):762-775. DOI: 10.1093/procel/pwad026

[4] Olorocisimo JP, Diaz LA, Co DE, Carag HM, Ibana JA, Velarde MC. *Lactobacillus delbrueckii* reduces anxiety-like behavior in zebrafish through a gut microbiome - Brain crosstalk. Neuropharmacology. 2023;**225**:109401. DOI: 10.1016/j.neuropharm.2022.109401. Epub 2022 Dec 21

[5] James DM, Davidson EA, Yanes J, Moshiree B, Dallman JE. The gut-brain-microbiome axis and its link to autism: Emerging insights and the potential of zebrafish models. Frontiers in Cell and Development Biology. 2021;**9**:662916. DOI: 10.3389/fcell.2021.662916

[6] Stagaman K, Alexiev A, Sieler MJ, Hammer A, Kasschau KD, Truong L, et al. The zebrafish gut microbiome influences benzo[a]pyrene developmental neurobehavioral toxicity. Scientific Reports. 2024;**14**(1):14618. DOI: 10.1038/s41598-024-65610-3

[7] Sireswar S, Dey G, Biswas S. Influence of fruit-based beverages on efficacy of *Lacticaseibacillus rhamnosus* GG (*Lactobacillus rhamnosus* GG) against DSS-induced intestinal inflammation. Food Research International. 2021;**149**:110661. DOI: 10.1016/j.foodres.2021.110661. Epub 2021 Aug 25

[8] Lozupone CA, Stombaugh JI, Gordon JI, Jansson JK, Knight R. Diversity, stability and resilience of the human gut microbiota. Nature. 2012;**489**(7415):220-230. DOI: 10.1038/nature11550

[9] Cryan JF, O'Riordan KJ, Cowan CSM, Sandhu KV, Bastiaanssen TFS, Boehme M, et al. The microbiota-gut-brain axis. Physiological Reviews. 2019;**99**(4):1877-2013. DOI: 10.1152/physrev.00018.2018

[10] Valenzuela MJ, Caruffo M, Herrera Y, Medina DA, Coronado M, Feijóo CG, et al. Evaluating the capacity of human gut microorganisms to colonize the zebrafish larvae (*Danio rerio*). Frontiers in Microbiology. 2018;**9**:1032. DOI: 10.3389/fmicb.2018.01032

[11] Davis DJ, Bryda EC, Gillespie CH, Ericsson AC. Microbial modulation of behavior and stress responses in zebrafish larvae. Behavioural Brain Research. 2016;**311**:219-227. DOI: 10.1016/j.bbr.2016.05.040. Epub 2016 May 20

[12] Phillips JB, Westerfield M. Zebrafish models in translational research: Tipping the scales toward advancements in human health. Disease Models & Mechanisms. 2014;7(7):739-743. DOI: 10.1242/dmm.015545

[13] Stewart AM, Braubach O, Spitsbergen J, Gerlai R, Kalueff AV. Zebrafish models for translational neuroscience research: From tank to

bedside. Trends in Neurosciences. 2014;**37**(5):264-278. DOI: 10.1016/j.tins.2014.02.011. Epub 2014 Apr 9

[14] Adhish M, Manjubala I. Effectiveness of zebrafish models in understanding human diseases-A review of models. Heliyon. 2023;**9**(3):e14557. DOI: 10.1016/j.heliyon.2023.e14557

[15] Ferranti EP, Dunbar SB, Dunlop AL, Corwin EJ. 20 things you didn't know about the human gut microbiome. The Journal of Cardiovascular Nursing. 2014;**29**(6):479-481. DOI: 10.1097/JCN.0000000000000166

[16] Latalova K, Hajda M, Prasko J. Can gut microbes play a role in mental disorders and their treatment? Psychiatria Danubina. 2017;**29**(1):28-30. DOI: 10.24869/psyd.2017.28

[17] Sender R, Fuchs S, Milo R. Are we really vastly outnumbered? Revisiting the ratio of bacterial to host cells in humans. Cell. 2016;**164**(3):337-340. DOI: 10.1016/j.cell.2016.01.013

[18] Roughgarden J. Holobiont evolution: Population theory for the Hologenome. The American Naturalist. 2023;**201**(6):763-778. DOI: 10.1086/723782. Epub 2023 Apr 17

[19] Carabotti M, Scirocco A, Maselli MA, Severi C. The gut-brain axis: Interactions between enteric microbiota, central and enteric nervous systems. Annals of Gastroenterology. 2015;**28**(2):203-209

[20] Hugon P, Dufour JC, Colson P, Fournier PE, Sallah K, Raoult D. A comprehensive repertoire of prokaryotic species identified in human beings. The Lancet Infectious Diseases. 2015;**15**(10):1211-1219. DOI: 10.1016/S1473-3099(15)00293-5. Epub 2015 Aug 23

[21] Li J, Jia H, Cai X, Zhong H, Feng Q, Sunagawa S, et al. An integrated catalog of reference genes in the human gut microbiome. Nature Biotechnology. 2014;**32**(8):834-841. DOI: 10.1038/nbt.2942. Epub 2014 Jul 6

[22] Eckburg PB, Bik EM, Bernstein CN, Purdom E, Dethlefsen L, Sargent M, et al. Diversity of the human intestinal microbial flora. Science. 2005;**308**(5728):1635-1638. DOI: 10.1126/science.1110591. Epub 2005 Apr 14

[23] Borrelli L, Aceto S, Agnisola C, De Paolo S, Dipineto L, Stilling RM, et al. Probiotic modulation of the microbiota-gut-brain axis and behaviour in zebrafish. Scientific Reports. 2016;**6**:30046. DOI: 10.1038/srep30046

[24] Teng M, Zhao X, Wang C, Wang C, White JC, Zhao W, et al. Polystyrene nanoplastics toxicity to zebrafish: Dysregulation of the brain-intestine-microbiota axis. ACS Nano. 2022;**16**(5):8190-8204. DOI: 10.1021/acsnano.2c01872. Epub 2022 May 4

[25] Shi Y, Chen C, Han Z, Chen K, Wu X, Qiu X. Combined exposure to microplastics and amitriptyline caused intestinal damage, oxidative stress and gut microbiota dysbiosis in zebrafish (*Danio rerio*). Aquatic Toxicology. 2023;**260**:106589. DOI: 10.1016/j.aquatox.2023.106589. Epub 2023 May 24

[26] Elinav E, Devkota S, Meisel M, Zhu S, Chu H, Chen H, et al. Microbes and metabolites in immunity. Immunity (Voices). 2024;**57**(9):1995-1999

[27] Spencer NJ, Hu H. Enteric nervous system: Sensory transduction, neural circuits and gastrointestinal motility. Nature Reviews. Gastroenterology & Hepatology. 2020;**17**(6):338-351. DOI: 10.1038/s41575-020-0271-2. Epub 2020 Mar 9

[28] Furness JB. The enteric nervous system and neurogastroenterology. Nature Reviews. Gastroenterology & Hepatology. 2012;**9**(5):286-294. DOI: 10.1038/nrgastro.2012.32

[29] Vicentini FA, Keenan CM, Wallace LE, Woods C, Cavin JB, Flockton AR, et al. Intestinal microbiota shapes gut physiology and regulates enteric neurons and glia. Microbiome. 2021;**9**(1):210. DOI: 10.1186/s40168-021-01165-z

[30] Ganz J, Melancon E, Eisen JS. Zebrafish as a model for understanding enteric nervous system interactions in the developing intestinal tract. Methods in Cell Biology. 2016;**134**:139-164. DOI: 10.1016/bs.mcb.2016.02.003. Epub 2016 Mar 9

[31] Qiao R, Sheng C, Lu Y, Zhang Y, Ren H, Lemos B. Microplastics induce intestinal inflammation, oxidative stress, and disorders of metabolome and microbiome in zebrafish. Science of The Total Environment. 2019;**662**:246-253. DOI: 10.1016/j.scitotenv.2019.01.245. Epub 2019 Jan 22

[32] Zhang B, Chen T, Cao M, Yuan C, Reiter RJ, Zhao Z, et al. Gut microbiota dysbiosis induced by decreasing endogenous melatonin mediates the pathogenesis of Alzheimer's disease and obesity. Frontiers in Immunology. 2022;**13**:900132. DOI: 10.3389/fimmu.2022.900132

[33] Naseribafrouei A, Hestad K, Avershina E, Sekelja M, Linløkken A, Wilson R, et al. Correlation between the human fecal microbiota and depression. Neurogastroenterology and Motility. 2014;**26**(8):1155-1162. DOI: 10.1111/nmo.12378. Epub 2014 Jun 1

[34] Hu S, Li A, Huang T, Lai J, Li J, Sublette ME, et al. Gut microbiota changes in patients with bipolar depression. Advanced Science (Weinheim, Baden-Württemberg, Germany). 2019;**6**(14):1900752. DOI: 10.1002/advs.201900752

[35] Góralczyk-Bińkowska A, Szmajda-Krygier D, Kozłowska E. The microbiota-gut-brain axis in psychiatric disorders. International Journal of Molecular Sciences. 2022;**23**(19):11245. DOI: 10.3390/ijms231911245

[36] Socała K, Doboszewska U, Szopa A, Serefko A, Włodarczyk M, Zielińska A, et al. The role of microbiota-gut-brain axis in neuropsychiatric and neurological disorders. Pharmacological Research. 2021;**172**:105840. DOI: 10.1016/j.phrs.2021.105840. Epub 2021 Aug 24

[37] Hyland NP, Cryan JF. Microbe-host interactions: Influence of the gut microbiota on the enteric nervous system. Developmental Biology. 2016;**417**(2):182-187. DOI: 10.1016/j.ydbio.2016.06.027. Epub 2016 Jun 22

[38] Rolig AS, Mittge EK, Ganz J, Troll JV, Melancon E, Wiles TJ, et al. The enteric nervous system promotes intestinal health by constraining microbiota composition. PLoS Biology. 2017;**15**(2):e2000689. DOI: 10.1371/journal.pbio.2000689

[39] Andersen-Civil AIS, Sawale RA, Vanwalleghem GC. Zebrafish (*Danio rerio*) as a translational model for neuro-immune interactions in the enteric nervous system in autism spectrum disorders. Brain, Behavior, and Immunity. 2023;**112**:254-266. DOI: 10.1016/j.bbi.2023.06.001. Epub 2023 Jun 9

[40] Agirman G, Yu KB, Hsiao EY. Signaling inflammation across the gut-brain axis. Science. 2021;**374**(6571):1087-1092. DOI: 10.1126/science.abi6087. Epub 2021 Nov 25

[41] Kamareddine L, Najjar H, Sohail MU, Abdulkader H, Al-Asmakh M. The microbiota and gut-related disorders: Insights from animal models. Cells. 2020;**9**(11):2401. DOI: 10.3390/cells9112401

[42] Levy M, Kolodziejczyk AA, Thaiss CA, Elinav E. Dysbiosis and the immune system. Nature Reviews. Immunology. 2017;**17**(4):219-232. DOI: 10.1038/nri.2017.7. Epub 2017 Mar 6

[43] Litvak Y, Bäumler AJ. The founder hypothesis: A basis for microbiota resistance, diversity in taxa carriage, and colonization resistance against pathogens. PLoS Pathogens. 2019;**15**(2):e1007563. DOI: 10.1371/journal.ppat.1007563

[44] Diaz OE, Castro RAM, Das S, Villablanca EJ. Experimental models of intestinal inflammation: Lessons from mouse and zebrafish. Molecular Genetics of Inflammatory Bowel Disease. 2019;**1**:47-76. DOI: 10.1007/978-3-030-28703-0_3

[45] Arias-Jayo N, Abecia L, Alonso-Sáez L, Ramirez-Garcia A, Rodriguez A, Pardo MA. High-fat diet consumption induces microbiota dysbiosis and intestinal inflammation in zebrafish. Microbial Ecology. 2018;**76**(4):1089-1101. DOI: 10.1007/s00248-018-1198-9. Epub 2018 May 7

[46] Azbazdar Y, Poyraz YK, Ozalp O, Nazli D, Ipekgil D, Cucun G, et al. High-fat diet feeding triggers a regenerative response in the adult zebrafish brain. Molecular Neurobiology. 2023;**60**(5):2486-2506. DOI: 10.1007/s12035-023-03210-4. Epub 2023 Jan 21

[47] Lee JG, Cho HJ, Jeong YM, Lee JS. Genetic approaches using zebrafish to study the microbiota-gut-brain axis in neurological disorders. Cells. 2021;**10**(3):566. DOI: 10.3390/cells10030566

[48] Lin W, Li K, Qin Y, Han X, Chen X, Ren Y. *Flunitrazepam* induces neurotoxicity in zebrafish through microbiota-gut-brain axis. Science of The Total Environment. 2023;**901**:165974. DOI: 10.1016/j.scitotenv.2023.165974. Epub ahead of print

[49] Wang Y, Song J, Wang X, Qian Q, Wang H. Study on the toxic-mechanism of triclosan chronic exposure to zebrafish (*Danio rerio*) based on gut-brain axis. Science of The Total Environment. 2022;**844**:156936. DOI: 10.1016/j.scitotenv.2022.156936. Epub 2022 Jun 27

[50] Zhong X, Li J, Lu F, Zhang J, Guo L. Application of zebrafish in the study of the gut microbiome. Animal Models and Experimental Medicine. 2022;**5**(4):323-336. DOI: 10.1002/ame2.12227. Epub 2022 Apr 12

[51] MacRae CA, Peterson RT. Zebrafish as tools for drug discovery. Nature Reviews. Drug Discovery. 2015;**14**(10):721-731. DOI: 10.1038/nrd4627. Epub 2015 Sep 11

[52] Holtzman NG, Iovine MK, Liang JO, Morris J. Learning to fish with genetics: A primer on the vertebrate model *Danio rerio*. Genetics. 2016;**203**(3):1069-1089. DOI: 10.1534/genetics.116.190843

[53] Garcia GR, Noyes PD, Tanguay RL. Advancements in zebrafish applications for 21st century toxicology. Pharmacology & Therapeutics. 2016;**161**:11-21. DOI: 10.1016/j.pharmthera.2016.03.009. Epub 2016 Mar 22

[54] Westerfield M. The Zebrafish Book. A Guide for the Laboratory Use of Zebrafish (*Danio rerio*). 5th ed. Eugene: University of Oregon Press; 2007

[55] Kimmel CB, Ballard WW, Kimmel SR, Ullmann B, Schilling TF.

Stages of embryonic development of the zebrafish. Developmental Dynamics. 1995;**203**(3):253-310. DOI: 10.1002/aja.1002030302

[56] He JH, Gao JM, Huang CJ, Li CQ. Zebrafish models for assessing developmental and reproductive toxicity. Neurotoxicology and Teratology. 2014;**42**:35-42. DOI: 10.1016/j.ntt.2014.01.006. Epub 2014 Feb 3

[57] McMenamin SK, Chandless MN, Parichy DM. Working with zebrafish at postembryonic stages. Methods in Cell Biology. 2016;**134**:587-607. DOI: 10.1016/bs.mcb.2015.12.001. Epub 2016 Feb 28

[58] Ng AN, de Jong-Curtain TA, Mawdsley DJ, White SJ, Shin J, Appel B, et al. Formation of the digestive system in zebrafish: III. Intestinal epithelium morphogenesis. Developmental Biology. 2005;**286**(1):114-135. DOI: 10.1016/j.ydbio.2005.07.013

[59] Wallace KN, Akhter S, Smith EM, Lorent K, Pack M. Intestinal growth and differentiation in zebrafish. Mechanisms of Development. 2005;**122**(2):157-173. DOI: 10.1016/j.mod.2004.10.009

[60] Brugman S. The zebrafish as a model to study intestinal inflammation. Developmental & Comparative Immunology. 2016;**64**:82-92. DOI: 10.1016/j.dci.2016.02.020

[61] Bates JM, Mittge E, Kuhlman J, Baden KN, Cheesman SE, Guillemin K. Distinct signals from the microbiota promote different aspects of zebrafish gut differentiation. Developmental Biology. 2006;**297**(2):374-386. DOI: 10.1016/j.ydbio.2006.05.006. Epub 2006 May 11

[62] Jijon H, Suarez-Lopez L, Diaz O, et al. Intestinal epithelial cell-specific RARα depletion results in aberrant epithelial cell homeostasis and underdeveloped immune system. Mucosal Immunology. 2018;**11**:703-715. DOI: 10.1038/mi.2017.91

[63] Katano T, Bialkowska AB, Yang VW. KLF4 regulates goblet cell differentiation in BMI1+ reserve intestinal stem cell lineage during homeostasis. International Journal of Stem Cells. 2020;**13**(3):424-431. DOI: 10.15283/ijsc20048

[64] Langenau DM, Ferrando AA, Traver D, Kutok JL, Hezel JP, Kanki JP, et al. In vivo tracking of T cell development, ablation, and engraftment in transgenic zebrafish. Proceedings of the National Academy of Sciences of the United States of America. 2004;**101**(19):7369-7374. DOI: 10.1073/pnas.0402248101. Epub 2004 May 3

[65] Wallace KN, Pack M. Unique and conserved aspects of gut development in zebrafish. Developmental Biology. 2003;**255**(1):12-29. DOI: 10.1016/s0012-1606(02)00034-9

[66] Rawls JF, Mahowald MA, Ley RE, Gordon JI. Reciprocal gut microbiota transplants from zebrafish and mice to germ-free recipients reveal host habitat selection. Cell. 2006;**127**(2):423-433. DOI: 10.1016/j.cell.2006.08.043

[67] Laing KJ, Purcell MK, Winton JR, Hansen JD. A genomic view of the NOD-like receptor family in teleost fish: Identification of a novel NLR subfamily in zebrafish. BMC Evolutionary Biology. 2008;**8**:42. DOI: 10.1186/1471-2148-8-42

[68] Howe K, Clark MD, Torroja CF, et al. The zebrafish reference genome sequence and its relationship to the human genome. Nature. 2013;**496**(7446):498-503. DOI: 10.1038/nature12111. Epub 2013 Apr 17

[69] Miklósi A, Andrew RJ. The zebrafish as a model for behavioral studies. Zebrafish. 2006;**3**(2):227-234. DOI: 10.1089/zeb.2006.3.227

[70] Qiao F, Tan F, Li LY, Lv HB, Chen L, Du ZY, et al. Alteration and the function of intestinal microbiota in high-fat-diet- or genetics-induced lipid accumulation. Frontiers in Microbiology. 2021;**12**:741616. DOI: 10.3389/fmicb.2021.741616

[71] Pei X, Heng X, Chu W. Polystyrene nano/microplastics induce microbiota dysbiosis, oxidative damage, and innate immune disruption in zebrafish. Microbial Pathogenesis. 2022;**163**:105387. DOI: 10.1016/j.micpath.2021.105387. Epub 2022 Jan 4

[72] Cao X, Chen L, Fan Y, Fu M, Du Q, Chang Z. Black phosphorus quantum dots induced neurotoxicity, intestinal microbiome and metabolome dysbiosis in zebrafish (*Danio rerio*). Science of The Total Environment. 2024;**954**:176644. DOI: 10.1016/j.scitotenv.2024.176644. Epub 2024 Oct 5

[73] Huang J, Wang Q, Liu S, Lai H, Tu W. Comparative chronic toxicities of PFOS and its novel alternatives on the immune system associated with intestinal microbiota dysbiosis in adult zebrafish. Journal of Hazardous Materials. 2022;**425**:127950. DOI: 10.1016/j.jhazmat.2021.127950. Epub 2021 Dec 2

[74] Melancon E, De La Torre G, Canny S, Sichel S, Kelly M, Wiles TJ, et al. Best practices for germ-free derivation and gnotobiotic zebrafish husbandry. Methods in Cell Biology. 2017;**138**:61-100. DOI: 10.1016/bs.mcb.2016.11.005

[75] Pham LN, Kanther M, Semova I, Rawls JF. Methods for generating and colonizing gnotobiotic zebrafish. Nature Protocols. 2008;**3**(12):1862-1875. DOI: 10.1038/nprot.2008.186

[76] Weitekamp CA, Kvasnicka A, Keely SP, Brinkman NE, Howey XM, Gaballah S, et al. Monoassociation with bacterial isolates reveals the role of colonization, community complexity and abundance on locomotor behavior in larval zebrafish. Animal Microbiome. 2021;**3**(1):12. DOI: 10.1186/s42523-020-00069-x

[77] Brinkmann BW, Koch BEV, Spaink HP, Peijnenburg WJGM, Vijver MG. Colonizing microbiota protect zebrafish larvae against silver nanoparticle toxicity. Nanotoxicology. 2020;**14**(6):725-739. DOI: 10.1080/17435390.2020.1755469

Chapter 5

Bridging Traditional and Modern Medicine: The Role of Zebrafish in *Ayurveda* Drug Discovery

Priya Gupta, Dayanand Hiremath, Laxmi Bhusal, Thakur Rakesh Singh and Anupam Srivastava

Abstract

Ayurveda has been an integral part of healthcare delivery systems for centuries in different countries of the world in one or another form, offering therapeutic benefits in many disease conditions. However, suboptimal details in pharmacokinetic studies and inadequate scientific validation on safety, efficacy, and the mode of action of Ayurvedic drugs remain a critical area of research. Recently, the zebrafish (*Danio rerio*) vertebrate model has gained prominence as a versatile and powerful tool for studying the pharmacological and toxicological effects of plant-based compounds. Due to its genetic similarity to humans, rapid development, tiny size, optical clarity of embryos, and suitability for high-throughput screening, zebrafish serves as an effective model for evaluating the efficacy and safety of natural product (plant/processed metal/mineral) formulations. The present chapter explores the integration of time-tested concepts of *Ayurveda*, especially its therapeutic potential, with zebrafish research, emphasizing their role in disease modeling and drug discovery. This chapter will focus on the application of zebrafish models for assessing polyherbal formulations in neurological disorders, metabolic diseases, and toxicological evaluations. Given the growing interest in natural product-based therapeutics and the limitations of conventional drug discovery, this chapter underscores the relevance of zebrafish models in bridging the conceptual gaps between the traditional Indian system of medicine and the conventional or modern system of medicine, which would be of significant interest to researchers exploring novel treatment strategies, contributing to advancing translational medicine and global healthcare.

Keywords: *Ayurveda*, zebrafish, Indian traditional system, neurodegenerative disorders, metabolic diseases

1. Introduction

Natural products, particularly plants/processed herb-mineral products, have played a vital role in maintaining human health and wellness for centuries. Rooted in ancient civilizations, herbal remedies have been widely utilized across diverse cultures to prevent and manage various ailments. There are several forms of traditional medicine, such as traditional Chinese medicine, *Ayurveda,* Kampo, traditional Korean medicine, Siddha, and Unani, that have been practiced for ages worldwide. However, one of the most comprehensive and enduring Indian traditional systems is *Ayurveda*, which focuses on promoting longevity and addressing human health conditions through a holistic approach. The *Ayurveda* term is derived from the Sanskrit words "*Ayu*" (meaning life or lifespan) and "*Veda*" (meaning knowledge or science), hence often translated as the "Science of Life." The origin of *Ayurveda* can be traced back to the Vedic period, when sages and seers recorded their knowledge of health and healing in sacred texts in the form of the *Vedas*. Among these, the *Atharva Veda* contains detailed descriptions of various diseases, their causes, symptoms, and treatments. Over time, *Ayurveda* evolved through the contributions of renowned scholars such as *Charaka* and *Sushruta*. Their works, the *Charaka Samhita* and *Sushruta Samhita*, are foundational texts that offer comprehensive guidelines for lifestyle, diagnosis, treatment, and disease prevention [1].

Ayurveda views the human body as composed of five fundamental elements: *Prithvi* (earth), *Tejas* (fire), *Vayu* (air), and *Akasha* (space/sky). These elements exist in varying ratios within the body as well as in the external universe, influencing both form and function. *Ayurveda* also recognizes the concepts of *Tridosha* (the three bodily humors: *Vata*, *Pitta*, and *Kapha*) and *Triguna* (the three mental qualities) as key frameworks for physical and psychological well-being. Its approach to healing is holistic, incorporating various concepts, dietary guidance, lifestyle modifications, and psychological support [2]. In the context of neurodegenerative diseases, *Ayurveda* provides a distinctive framework that emphasizes systemic imbalances and holistic health. The *Tridosha* system plays a central role, with *Vata* regulating movement and communication, *Pitta* managing metabolism and energy, and *Kapha* supporting growth and structural integrity. Any kind of imbalances in these energies, often triggered by poor dietary habits or environmental factors, can lead to the onset of disease [3]. *Ayurveda* utilizes herbal remedies and therapeutic procedures like *Nasya*, *Basti,* and *Shirodhara* to treat neurological conditions linked to *Vata* disorders. Emerging scientific evidence supports its effectiveness, with studies highlighting the neuroprotective benefits of herbs such as *ashwagandha* and *brahmi* in both animal and clinical trials. Clinical reasoning is found guided by systematic observation and rational analysis (*yukti*), employing retrospective (*purvavat*), prospective (*sesavat*), and cross-sectional (*samanyatodrsta*) methodologies. Treatment outcomes are critically evaluated by comparing the results of intervention (*paksa*) against both positive (*sapaksa*) and negative (*vipaksa*) controls to establish causality [4].

Furthermore, Ayurvedic medicine treats various physical and mental health conditions using natural ingredients derived from plants, minerals, and animal sources to a certain extent. In *Ayurveda*, medicinal herbs are utilized in various forms and formulations to treat diseases and promote overall well-being. The herbs can be used individually or combined into formulations depending on the nature and properties of the herbs, the condition being treated, and the patient's body constitution. According to the World Health Organization (WHO), herbal medicines are comprised of herbs, herbal materials, herbal preparations, and finished herbal products that are formulated in suitable forms for administration to patients. Approximately 80% of the global population, particularly in developing countries, relies on herbal medicine as part of their primary healthcare management [5], highlighting the ongoing relevance and importance of traditional medicinal systems in current times.

These medicines, derived either from a single plant or a combination of multiple herbs, are typically processed into various dosage forms such as tablets, capsules, syrups, decoctions, medicated oil, powders, and more.

In recent years, the demand for *Ayurvedic*-derived products, often formulated from plant-based extracts, has been growing steadily due to many reasons, such as increased global interest in natural and traditional remedies, available research leads, and expensive research discoveries. However, despite their widespread use, the scientific research on phytochemical analysis, pharmacological studies, and preclinical and clinical trials in herbal medicines remains limited. This lack of in-depth study creates challenges in understanding their therapeutic potential, mechanism of action, and safety profiles. Thus, the scientific validation of plant medicines through standardization, purity evaluation, and the demonstration of efficacy in both animal models and human studies is essential for enhancing credibility and increasing their acceptance in modern healthcare [6].

Zebrafish (*Danio rerio*) have emerged as a widely used model organism in contemporary biomedical research. Native to tropical freshwater habitats, zebrafish are commonly found in South Asian countries like India, Nepal, and Bangladesh [7]. The zebrafish life cycle comprises four distinct stages: embryo, larva, juvenile, and adult. Under favorable conditions, a single zebrafish can lay up to 200 eggs per week, making them highly efficient for large-scale breeding and experimental studies. Due to several unique advantages, zebrafish are increasingly being used as an alternative to traditional mammalian models for studying human diseases and genetics [8]. Genetically, zebrafish share remarkable similarities with humans—approximately 70% of human protein-coding genes have orthologs in zebrafish, and more than 80% of human disease-associated genes have identifiable counterparts in this model organism [9]. Therefore, zebrafish have become an important model in a wide range of biomedical studies, including reproductive studies, neurological disorders, cardiovascular diseases, oncology, and drug discovery [4].

2. Zebrafish as a preclinical model for herbal/herb-mineral-based formulations

Zebrafish have emerged as a promising and valuable model in drug discovery [10], and their application is gaining momentum in the preclinical evaluation of herbal/herb-mineral-based formulations. Their unique biological attributes, such as rapid embryonic development, optical transparency, external fertilization, and high genetic homology to humans, make them especially valuable for studying developmental, neurobehavioral, toxicological, and pharmacological responses to traditional medicines. As *Ayurveda* continues to be widely practiced in several regions worldwide, the need for scientific validation of its formulations has become increasingly important for global acceptance. Zebrafish models enable researchers to conduct early-phase investigations into the safety and therapeutic efficacy of both plant-based and herbo-mineral Ayurvedic preparations. Their small size and amenability to high-throughput screening allow for cost-effective and ethically favorable experimentation. In *Ayurveda*-inspired research, zebrafish are commonly used for assessing acute and chronic toxicity, teratogenicity, behavioral outcomes, and organ-specific effects. The transparent embryos provide a real-time window into developmental processes, making them ideal for visualizing malformations, heart rate changes, and organogenesis following treatment with herbal or metallic formulations. Notably, studies have shown that *Kajjali—Black Sulphide of Mercury* (tested up to 20 ppm for 48 hours) and *Rasasindura—Red Sulphide of Mercury* (up to 75 ppm for 48 hours) exhibit significantly lower toxicity compared to conventional mercury compounds like mercuric chloride ($HgCl_2$), methylmercury, or elemental mercury (Hg), when prepared according to classical Ayurvedic methods [11]. Additionally, adult zebrafish serve as effective models for behavioral assays related to anxiety, cognition, locomotion, and social interaction parameters often targeted by Ayurvedic therapies. For instance, *Suvarna Bhasma* (Gold Ash), an Ayurvedic preparation, revealed no adverse effects on zebrafish motor functions, with treated fish demonstrating behavior similar to that of the control group. Interestingly, Swarna Bhasma (SB) also exhibited anxiolytic properties, as evidenced by increased time spent in the upper zone of the tank and elevated average swimming height. Remarkably, these behavioral improvements persisted for more than 30 days after treatment cessation [12].

Zebrafish have also been instrumental in evaluating the mechanisms of action of Ayurvedic compounds through tools such as *in vivo* imaging, gene expression analysis, and omics technologies. These capabilities allow for the investigation of complex pharmacological pathways and potential drug targets. For example, studies have utilized zebrafish to understand how certain herbal extracts modulate oxidative stress, inflammation, or neurotransmitter systems' core mechanisms linked with *Ayurveda*'s therapeutic strategies. Expanding the scope to high-throughput screening applications, zebrafish have also been used to assess the prophylactic anti-inflammatory potential of Ayurvedic formulations such as Patanjali special *chyawanprash*. In models where inflammation was induced using lipopolysaccharide (LPS) to simulate hypoxia-related conditions, *chyawanprash* was found to mitigate the resulting increase in heartbeat. Moreover, it effectively alleviated inflammation-induced locomotor impairments in zebrafish, underscoring its potential as a natural therapeutic [13]. Overall, the integration of zebrafish as an experimental model offers a translational platform to bridge traditional Ayurvedic knowledge with contemporary biomedical science. It enables the validation of safety and efficacy profiles while also uncovering the molecular underpinnings of therapeutic action. This makes zebrafish an invaluable tool in advancing *Ayurveda*-inspired preclinical research toward globally acceptable evidence-based medicine.

3. Plant-based compounds and their pharmacological evaluation in zebrafish

3.1 Scientific and mechanistic validation of ayurvedic formulations using the zebrafish model

Despite centuries of traditional use, many Ayurvedic-derived products still lack robust evidence regarding their pharmacokinetics, therapeutic efficacy, and safety. This is particularly important in regions where these formulations are commonly used without standardized validation. Hence, integrating modern tools like phytochemical profiling, pharmacological assays, and zebrafish-based preclinical models is crucial for bringing these treatments into mainstream healthcare. Zebrafish offer a scalable *in vivo* platform to simultaneously assess toxicity, behavioral effects, developmental impact, and underlying molecular mechanisms. For instance, the aqueous extract of *Bacopa monnieri*, traditionally used for cognitive enhancement, exhibited strong antioxidant properties with minimal developmental toxicity in zebrafish embryos at lower concentrations, suggesting its potential as a prophylactic agent against oxidative stress-related diseases. The phytochemical analysis confirmed the presence of flavonoids and polyphenolics, and antioxidant assays such as diphenyl-1-picrylhydrazyl (DPPH), azinobis[3-ethylbenzothiazoline-6-sulfonate] (ABTS), and ferric reducing antioxidant power (FRAP) revealed high free radical scavenging potential [14]. Similarly, *Benincasa hispida* fruit aqueous extract, traditionally used to treat mental disorders, demonstrated anxiolytic effects *via* behavioral assays like the light–dark preference and T-maze tests. Docking studies confirmed the involvement of active compounds in AChE inhibition, offering mechanistic insight into its neuropharmacological effects [15]. Divya-Herbal-Peya (DHP), a plant-based decoction usually prescribed by Ayurvedic practitioners, was phytochemically characterized and shown to contain 14 active Ayurvedic compounds like glycyrrhizin, rosmarinic acid, gingerol, cinnamic acid, protocatechuic acid, gallic acid, ellagic acid, piperine, and cinnamaldehyde. In LPS-induced zebrafish, DHP significantly prevented behavioral fever, locomotor changes, and fin damage and demonstrated upregulation of *TNFα, IL6,* and *IL1b*. These findings experimentally validate DHP's immunomodulatory potential in infection-related inflammatory responses [13, 16]. These examples underscore the importance of combining phytochemical characterization with functional assays and mechanistic studies, using zebrafish as a cost-effective preclinical model. However, further translational studies, including rodent models and clinical trials, are required to confirm long-term safety and therapeutic potential.

Furthermore, zebrafish have proven highly effective in elucidating mechanisms of action through advanced tools such as network pharmacology, transcriptomics, and molecular docking. For instance, *Chenopodium album* L. extract reduced inflammation in zebrafish by downregulating key pro-inflammatory signaling genes, including *PI3K, Akt, MAPK1,* and *NFκB*, while concurrently upregulating the anti-inflammatory PPAR pathway, as confirmed by transcriptomics and real-time quantitative polymerase chain reaction (RT-qPCR). These findings suggest a multi-targeted mechanism by which *C. album* may exert its therapeutic effects in inflammation modulation [17]. Similarly, turmeric leaf extract exposed to zebrafish embryos demonstrated strong antioxidant potential by reducing reactive oxygen species (ROS), lipid peroxidation, and cell death in the H_2O_2-induced oxidative stress model [18]. In a similar vein, another study was conducted using two zebrafish inflammation models—tail fin injury and LPS-induced intestinal inflammation—to examine the anti-inflammatory effects of *Urtica cannabina* L. (UL). In the tail injury model, UL at 10 mg/g significantly promoted fin regeneration and reduced scarring, whereas a higher dose of 30 mg/g was less effective. In the intestinal inflammation model, LPS treatment reduced feed intake, body weight, and goblet cell numbers while increasing the expression of pro-inflammatory genes (*TNFα, IL6, and IL8*). These adverse effects were significantly reversed with both 10 and 30 mg/g UL treatments, demonstrating its protective and reparative effects. Additionally, UL supplementation modulated gut microbiota by increasing beneficial *Bacteroidetes* and reducing harmful bacterial populations [19, 20].

Moreover, by integrating advanced approaches like network pharmacology and transcriptomics, several studies have revealed how bioactive compounds from medicinal herbs modulate key cellular pathways. Molecular docking is a valuable tool in addressing the complex task of identifying individual metabolites within herbal extracts, aiding in the chemical characterization and prediction of bioactivity. Numerous studies have employed this technique to isolate herbal bioactives and evaluate their therapeutic potential across various disorders [21]. In one such study, molecular docking combined with a zebrafish model was used to investigate the anti-inflammatory activity of *Boswellia serrata* extract. *In silico* analyses, including ADMET profiling, identified acetyl-11-keto-boswellic acid and 11-keto-β-boswellic acid as key constituents, both showing high binding affinity toward inflammation-specific receptors and favorable oral bioavailability. *In vivo*, the extract exhibited protective effects against LPS-induced inflammation in zebrafish larvae at concentrations ranging from 10 to 80 μg/mL. Specifically, treatment with 80 μg/mL significantly reduced intracellular reactive oxygen species (ROS), apoptosis, lipid peroxidation, and nitric oxide levels. Furthermore, the extract suppressed the expression of key pro-inflammatory genes, such as *iNOS, TNFα, COX2,* and *IL1*, indicating potent anti-inflammatory activity [22]. In another study, the anxiolytic and antidepressant effects of *Tribulus terrestris* (Tt) ethanolic extract were investigated using a zebrafish model of scopolamine-induced amnesia, supported by molecular docking studies targeting the monoamine oxidase A (MAO-A) enzyme. MAO-A plays a vital role in the metabolism of neurotransmitters in both the central nervous system and peripheral tissues and is known to be involved in alleviating depression and specific types of anxiety. In the study, six flavonoid compounds—rutin, apigetrin, kaempferol, quercetin, luteolin, and epigallocatechin—along with six saponin compounds, such as saponin C, protodioscin, terrestrosin C, trillarin, terreside B, and disogluside, were subjected to molecular docking analysis against the MAO-A enzyme. Behavioral assessments, including the Novel Tank Diving Test (NTT) and Novel Approach Test (NAT), were used to evaluate the anxiolytic and antidepressant-like effects of Tt in scopolamine-treated zebrafish. The results demonstrated that administration of Tt significantly reduced anxiety- and depression-like behaviors. Molecular docking further revealed that four flavonoids—kaempferol, quercetin, luteolin, apigetrin, and epigallocatechin—exhibited strong interactions with the MAO-A enzyme, suggesting a possible mechanism for the observed effects. In contrast, rutin and all six tested saponins showed no significant binding activity toward MAO-A. These findings support the therapeutic potential of *Tribulus terrestris* in managing neuropsychiatric conditions through its active flavonoid constituents [23].

Together, these examples highlight a multitiered approach to validate Ayurvedic drugs beginning with phytochemical and antioxidant profiling, moving through behavioral and toxicological screening in zebrafish, and culminating in mechanistic elucidation *via in silico* tools and omics technologies. This integrated strategy offers a comprehensive framework for modernizing and validating traditional Ayurvedic knowledge.

3.2 Pharmacological evaluation

3.2.1 Neurodegenerative disorders

Neurons play a fundamental role in brain function, serving as the functional unit of communication within the nervous system. While most neurons originate in the brain, they are distributed throughout the body, forming an extensive network essential for sensory, motor, and cognitive processes. During early development, neural stem cells give rise to the majority of neurons, but this neurogenesis is markedly reduced in adulthood. Although neurons are long-lived, they are not immortal, and their progressive loss, whether in number, structure, or function, is termed neurodegeneration. This phenomenon is a core feature in the pathophysiology of numerous brain disorders and represents a major public health challenge. Neurodegeneration is often characterized by the dysfunction of synapses and neural networks and the accumulation of misfolded or abnormally modified protein aggregates in the brain. These changes impair normal neural communication and lead to the deterioration of cognitive and motor functions. Disorders in which neurodegeneration is the central hallmark are collectively referred to as neurodegenerative diseases (NDs). Prominent examples of NDs include Alzheimer's disease, Parkinson's disease (PD), Huntington's disease (HD), amyotrophic lateral sclerosis (ALS), motor neuron disease, prion diseases, spinal muscular atrophy, and spinocerebellar ataxias. These conditions, though distinct in their clinical presentations and underlying molecular mechanisms, share common themes of progressive neuronal loss and disrupted brain circuitry [24]. People often seek the best possible treatments to alleviate these conditions; however, there is currently no permanent cure for ND. As a result, there is growing public interest in plant-based or traditional medicines.

PD, primarily a neurodegenerative motor disorder, is characterized by resting tremor, bradykinesia, rigidity, freezing, and postural instability due to dopaminergic neuron loss and Lewy body formation [25]. James Parkinson in the eighteenth century initially described the disease as "shaking palsy," which was later named PD. However, the descriptions of equivalent Parkinsonian symptoms had already been recognized centuries earlier in the traditional Indian medical system of *Ayurveda*. As early as 300 BC, the condition referred to as "*Kampavata*" was described in classical Ayurvedic texts like the *Charaka Samhita*, which detailed symptoms such as head tremors (*Sirakampa*) and generalized body tremors, offering one of the earliest coherent accounts of Parkinsonian features. The primary symptoms of *Kampavata* are *Kampa* (tremor) and *Stambha* (rigidity or stiffness), which could be *Ekanga* (localized) or *Sarvanga* (generalized) and were regarded as abnormal patterns of the *Cala* (moving), a property of *Vata* humor [26]. Evidence has demonstrated that various traditional Ayurvedic medicines are used to treat PD [27].

Moreover, zebrafish are increasingly recognized as a valuable vertebrate model for studying PD due to their genetic resemblance to humans and suitability for molecular and imaging studies, offering a cost-effective alternative to rodent models. Despite their utility, research has so far explored only a few plant-based treatments, specifically *Alpinia oxyphylla*, *Bacopa monnieri*, *Canavalia gladiata*, *Centella asiatica*, *Paeonia suffruticosa*, and *Stachytarpheta indica,* with only *B. monnieri* and *C. asiatica* showing promising anti-Parkinsonian effects in zebrafish [28]. For instance, the hydro-alcoholic extract of *Centella asiatica* (*Brahmi*) in streptozotocin-induced memory dysfunction in adult zebrafish [29]. *Swarna Bhasma* (SB) exhibits neuroprotective effects in a rotenone-induced Parkinson's-like zebrafish model by preserving dopamine levels and normalizing behavioral and proteomic alterations. Differential expression of Parkinson's-related proteins was prevented in SB-treated groups, highlighting its therapeutic potential. SB has also demonstrated anxiolytic effects in zebrafish models [12]. 1-methyl-4-phenyl-1,2,3,6-tetrahydropyridine (MPTP), a neurotoxicant treatment in zebrafish, mimics PD pathology with dopaminergic neuron loss and motor impairment, reversible by deprenyl [30]. These genetic and toxin-induced zebrafish models enable mechanistic studies and drug screening for both familial and idiopathic PD. Not only compound formulations, but even single plant medicine is also used to treat ND; for instance, in PD, *Mucuna pruriens* (*Lajjalu*) extract, which contains approximately 40% levodopa—a key active compound widely used in the treatment of PD [31]. In chronic neurodegenerative conditions where memory loss is a major concern, *Shilajatu (SJ)* (Black Bitumen) has shown promising results. SJ not only improves memory but also helps reduce oxidative stress and symptoms of amnesia [32].

Alzheimer's disease is a progressive neurodegenerative disorder primarily characterized by the accumulation of amyloid-beta (Aβ) plaques and neurofibrillary tangles composed of hyperphosphorylated tau protein. One of the hallmark biochemical features of AD is a significant reduction in acetylcholine levels in the brain, which has been linked to Aβ accumulation. The aggregation of Aβ proteins disrupts synaptic transmission and initiates chronic inflammatory responses, contributing to neuronal dysfunction and cognitive decline. Quercetin, the most potent flavonoid, exhibits neuroprotective effects in an aluminum chloride-induced zebrafish model of AD by reducing oxidative stress and enhancing cognitive function. Computational and docking analyses further support its multitarget binding potential, highlighting its promise as a therapeutic agent [33]. Similarly, methanolic extract of *Salacia fruticosa* improved cognitive function and reduced oxidative stress in a zebrafish model of scopolamine-induced Alzheimer's disease. Phytochemical and docking studies revealed strong AChE inhibition and antioxidant potential, supporting its role as a neuroprotective agent.

3.2.2 Metabolic diseases (MD)

Metabolic diseases are those conditions that disrupt normal metabolism, the process by which the body converts food into energy and performs essential functions such as breathing, digestion, and cellular repair. These disorders often involve dysfunctions in enzymes, hormones, or organs that are key to metabolic regulation. Some of the major health conditions that arise as a consequence of metabolic diseases include cardiovascular diseases, obesity, and diabetes mellitus. These chronic conditions often require lifelong medication, which can, over time, lead to additional health complications. This has prompted growing public interest in plant-based medicines, which are considered safer for long-term use and have shown promising results in managing metabolic disorders. For instance, *Syzygium cumini* (Jambu) seeds have demonstrated significant effectiveness against diabetes mellitus and hyperlipidemic conditions [34]. Similarly, the antidiabetic activity of *Vitex trifolia* bark has been documented in zebrafish models, where the bioactive compounds present in the plant were found to enhance insulin sensitivity [35]. Furthermore, plant extracts from *Urtica dioica, Apium graveolens,* and *Zingiber officinale* have been shown to exert hypoglycemic and hypolipidemic effects. These effects are achieved through the reduction of liver weight and improvement in the activities of key hepatic enzymes, including fructose-1,6-bisphosphatase, hexokinase, glucose-6-phosphatase, and phosphofructokinase, as well as alkaline phosphatase (ALP), alanine aminotransferase (ALT), and aspartate aminotransferase (AST) [36]. In addition to diabetes and obesity, cardiovascular disease is another major concern strongly associated with metabolic disorders. In such conditions, plant-based therapies have emerged as vital alternatives due to their efficacy and lower risk of adverse effects, even with prolonged use. For example, turmeric extract has been proven to exhibit anti-atherosclerotic and hypolipidemic activities in zebrafish models [37], further highlighting the therapeutic potential of natural products in managing metabolic complications.

3.2.3 Reproductive study

Infertility (*Vandhyatva*) has become a significant health concern in the modern era, with contributions from both genders being nearly equal. In men, infertility often results from abnormalities in sperm structure, number, and function, a condition that can be correlated with *Klaibya* as described in *Ayurveda*. According to the WHO, infertility has been classified as a disability and a "disease of the reproductive system." In women, infertility may arise from abnormalities in reproductive structures such as ovulation disorders (*Beeja Dosha*) and fallopian tube blockages. According to *Ayurveda*, before planning for conception, it is essential for both partners to undergo *Shodhana* (body purification) therapies. This process is intended to cleanse the body of any *Dosha* imbalances, ensuring a healthy *Kshetra* (uterus) and *Beeja* (sperm and ovum) to facilitate successful conception [38]. To enhance reproductive health and sexual vitality in both genders, *Ayurveda* emphasizes the use of specific medicinal plants.

Zebrafish are highly suitable for reproductive studies due to their short reproductive cycle and strong hormonal and neuronal similarities to the human reproductive system. A well-established method for inducing fertility-related mutations involves exposing zebrafish to N-ethyl-N-nitrosourea (ENU) *via* water, enabling efficient screening of individuals with partial or complete sterility. Based on these advantages, zebrafish can be used as an excellent model organism to assess reproductive health. For instance, research has shown that the extract of *Bryonopsis laciniosa* (Shivlingi) seeds can restore fertility in zebrafish models by reviving cytological and anatomical features [39]. Similarly, certain pathological conditions, such as ovarian cysts and anovulation, significantly impair fertility. In such cases, treatment with Putrajeevak (*Putranjiva roxburghii*) seed oil has been proven effective. It increases the ovarian follicle count and the number of mature eggs while reducing the presence of ovarian cysts. Furthermore, in males, *Putrajeevak* enhances sperm count and improves sperm motility [40]. Polycystic Ovarian Syndrome (PCOS), one of the leading causes of infertility in women, can also be managed through Ayurvedic interventions. Treatment with *Saraca asoca* has shown promising results in reversing PCOS pathology, thus restoring fertility potential [41]. Thus, *Ayurveda* offers a holistic approach to infertility by emphasizing purification, restoring Dosha balance, and utilizing potent herbal therapies to rejuvenate the reproductive system in both men and women.

3.2.4 Hepatoprotective activity and hepatotoxicity

The liver is the largest glandular organ in vertebrates and plays a central role in metabolizing carbohydrates, lipids, proteins, hormones, and vitamins, while also functioning as a primary site for detoxification [42–44]. Hepatocytes contain metabolic enzymes located in the microsomes, mitochondria, and cytosol, which participate in the biotransformation of drugs and toxins *via* oxidation, reduction, and hydrolysis reactions primarily mediated by cytochrome P450 monooxygenase enzymes [43, 45]. Given these functions, the liver is particularly vulnerable to toxic contaminants. Hepatotoxicity, which refers to liver injury caused by toxic compounds, often results from pharmaceuticals or drugs that trigger oxidative stress and programmed cell death [46, 47]. In early developmental stages, zebrafish larvae are transparent, allowing direct visualization of internal organs, including the liver, under light microscopy, eliminating the need for invasive procedures [48–50]. Importantly, zebrafish hepatogenesis follows a developmental trajectory similar to that of mammals, particularly mice [51], and their liver exhibits structural and functional homology to human liver tissue [52]. This has made zebrafish a widely accepted preclinical model for evaluating liver toxicity [53, 54]. For instance, North et al. [55] successfully established a zebrafish model for acetaminophen-induced liver injury and identified compounds that enhanced the effect of N-acetylcysteine, a known antidote. Similarly, Zhang et al. [56, 57] utilized a transgenic zebrafish line with fluorescently labeled liver cells to screen various hepatotoxic agents.

Hepatotoxicity refers to liver damage caused by factors such as alcohol consumption, infections, and medications, which can further lead to various liver anomalies like fatty liver, jaundice, and cirrhosis. In such conditions, alongside treating the underlying diseases, it is equally important to focus on regenerating the damaged liver cells. In this regard, *Ayurveda* plays a significant role. It has been proven that formulations containing *Boerhavia diffusa*, *Phyllanthus niruri*, and *Solanum nigrum* are effective in reversing hepatocellular toxicity (HCT) in zebrafish models [58]. Furthermore, for symptomatic treatment, extracts of *Picrorhiza kurroa* have demonstrated effectiveness against steatohepatitis [59]. In addition to these, the extract of *Tinospora cordifolia* has shown hepatoprotective activity in post-HCT conditions [60]. Not only plant-based medicines but also mineral-based formulations play a crucial role in hepatoprotection. For example, *Dhatri Lauha* has exhibited hepatoprotective effects [61], and *Somanathi Tamra Bhasma* has also been found effective against hepatocellular toxicity [62]. Thus, the combined use of herbal and mineral Ayurvedic formulations offers a promising approach for both the management and regeneration of liver tissues affected by hepatotoxicity.

3.2.5 Cardiovascular disease and inflammatory diseases

Cardiovascular disease (CVD) refers to a group of disorders affecting the heart and blood vessels, including conditions such as coronary artery disease, heart failure, arrhythmias, and stroke. The main triggering factors for CVD include atherosclerosis, diabetes mellitus, and hypertension. In such conditions, Ayurvedic medicines have been proven to play a crucial role in management and prevention. For instance, Chyawanprash has been shown to regulate immune responses in a zebrafish model of inflammation by moderating inflammatory biomarkers [13, 16]. Similarly, in the context of CVD, Yogendra Ras has demonstrated effectiveness against cardiac hypertrophy induced in zebrafish models [63]. Not only mineral-based formulations but also plant-derived medicines have shown promising results. An extract of *Murraya koenigii* (curry leaf) has demonstrated significant biological activity as an angiogenesis suppressor in a zebrafish model [64]. Additionally, aqueous extracts of turmeric (*Curcuma longa*) and laurel have exhibited notable anti-atherosclerotic and hypolipidemic effects in zebrafish models [37]. Furthermore, trending Ayurvedic formulations like *Divya Arjun Kwath*, prepared from *Terminalia arjuna*, have shown positive responses against hypertrophy in murine cardiomyocytes through the modulation of oxidative stress, as validated in zebrafish models [65]. Collectively, these studies highlight the potential of Ayurvedic formulations, both mineral and plant-based, in mitigating various pathological factors associated with CVD, thereby offering integrative and evidence-backed therapeutic strategies.

4. Harnessing new technologies to advance traditional knowledge for global health

The integration of traditional and indigenous knowledge with modern scientific tools is opening new frontiers in health and environmental well-being. Rapid technological advancements, particularly in fields such as artificial intelligence (AI), neuroimaging, and pharmacology, are transforming how traditional medicine is studied and validated. AI, with its robust data-mining and pattern-recognition capabilities, is revolutionizing the way centuries-old healing knowledge is analyzed, enabling researchers to uncover meaningful trends and correlations across vast ethnomedical datasets. Similarly, neuroimaging techniques like functional magnetic resonance imaging (fMRI) have been instrumental in understanding the physiological effects of traditional practices such as yoga and meditation, providing measurable evidence of their benefits in stress reduction and mental health enhancement. In pharmacological research, approaches like ethnopharmacology and reverse pharmacology are guiding the discovery of novel therapeutic compounds by tracing traditional remedies back to their bioactive principles. This convergence of traditional wisdom and cutting-edge science is not only fostering innovation but also building a robust evidence base to support the global use of traditional medicine. As more individuals turn to traditional products and practices, the scientific community is increasingly invested in discerning what works, why it works, and how best to integrate it into modern healthcare systems.

5. Limitations/challenges and future perspectives

Bridging the gap between the Indian traditional system and the conventional system of medicine presents both significant opportunities and notable challenges. A major hurdle lies in the limited standardized scientific evidence supporting many Ayurvedic treatments concerning the pharmacodynamics, pharmacokinetics, and safety profiles of herbal formulations. Despite centuries of empirical use, many Ayurvedic medicines have not undergone the rigorous testing required by modern biomedical standards. Questions regarding the precise mechanism of action, therapeutic dosage, safety margins, and potential herb-drug interactions remain largely unanswered. Additionally, the presence of contaminants such as heavy metals or adulterants in poorly regulated herbal preparations further raises concerns about their safe use in combination with pharmaceutical drugs [60].

To bridge this gap, there is an urgent need for robust pharmacological and toxicological studies that can scientifically evaluate the efficacy and safety described in *Ayurveda.* Investigating the pharmacokinetics and pharmacodynamics of bioactive compounds is essential. Many herbs used in *Ayurveda* are rich sources of diverse phytochemicals, including alkaloids, flavonoids, terpenoids, coumarins, phenolics, nitrogenous compounds, and organosulfur compounds. These compounds have been shown to exert a wide range of biological activities, such as anti-inflammatory, immunomodulatory, antioxidant, anticancer, antimicrobial, antiproliferative, and neuroprotective effects [66]. For instance, compounds like curcumin from *Curcuma longa* have been studied for their antioxidant, anti-inflammatory, and anti-atherosclerotic properties [67]. The zebrafish model, combined with high-throughput screening (HTS), molecular docking, and bioinformatics, can be effectively employed to identify potential drug candidates from Ayurvedic herbs [68].

In conclusion, the path toward effective integration of *Ayurveda* with the conventional system of medicine lies in embracing a multidisciplinary approach, one that combines the holistic, preventive philosophy of *Ayurveda* with the analytical rigor of biomedical science. By investing in pharmacological research, preclinical studies, and advanced analytical techniques, it is possible to scientifically validate Ayurvedic treatments, optimize herbal formulations, and ensure their safe and effective use alongside conventional therapies. This integrative approach holds great promise in addressing chronic and lifestyle-related diseases, offering personalized, holistic, and sustainable healthcare solutions.

Acknowledgements

All authors thank the National Institute of Ayurveda, Jaipur, and the AYUSH government for providing the necessary resources and support to write the present chapter. This work received no external funding from any agency.

Author contributions

All authors listed have made a substantial contribution to this chapter.

Conflict of interest

The author declares that there is no conflict of interest.

Author details

Priya Gupta, Dayanand Hiremath, Laxmi Bhusal, Thakur Rakesh Singh and Anupam Srivastava*
Department of Rasashastra and Bhaishajya Kalpana, National Institute of Ayurveda, Jaipur, India

*Address all correspondence to: dras2008@gmail.com

References

[1] Verma SK, Pandey M, Sharma A, Singh D. Exploring ayurveda: Principles and their application in modern medicine. Bulletin of the National Research Centre. 2024;**48**(1):77. DOI: 10.1186/s42269-024-01231-0

[2] Kapur M. Basic principles of ayurveda. In: Psychological Perspectives on Childcare in Indian Indigenous Health Systems. India: Springer; 2016. pp. 15-29. DOI: 10.1007/978-81-322-2428-0_2

[3] Dubey A, Ghosh NS, Singh R. Zebrafish as an emerging model: An important testing platform for biomedical science. Journal of Pharmaceutical Negative Results. 2022;**13**(3):1-7. DOI: 10.47750/pnr.2022.13.03.001

[4] Ramachandran A, Patel NJ. Ayurveda after age of Covid-19 exploring the impact and empowering informed decisions through health literacy. 2024

[5] Bodeker G, Ong CK, Grundy C, Burford G, Shein K. WHO Global Atlas of Traditional, Complementary and Alternative Medicine. Kobe, Japan: World Health Organization (WHO); 2005

[6] Bhosale VV, Banerjee D. Scientific validation of herbal medicine. In: Herbal Medicine in India. Singapore: Springer; 2020. pp. 573-579. DOI: 10.1007/978-981-13-7248-3_34

[7] Rahman Khan F, Sulaiman Alhewairini S. Zebrafish (Danio rerio) as a model organism. In: Current Trends in Cancer Management. London: IntechOpen; 2019. DOI: 10.5772/intechopen.81517

[8] Cassar S, Adatto I, Freeman JL, Gamse JT, Iturria I, Lawrence C, et al. Use of zebrafish in drug discovery toxicology. Chemical Research in Toxicology. 2020;**33**(1):95-118. DOI: 10.1021/acs.chemrestox.9b00335

[9] Poureetezadi SJ, Donahue EK, Wingert RA. A manual small molecule screen approaching high-throughput using zebrafish embryos. Journal of Visualized Experiments: JoVE. 2014;**93**:52063

[10] Malicki J, Jo H, Wei X, Hsiung M, Pujic Z. Analysis of gene function in the zebrafish retina. Methods. 2002;**28**(4):427-438

[11] Biswas S, Bellare J. Explaining Ayurvedic preparation of Rasasindura, its toxicological effects on NIH3T3 cell line and zebrafish larvae. Journal of Ayurveda and Integrative Medicine. 2022;**13**(2):100518

[12] Biswas S, Dhumal R, Selkar N, Bhagat S, Chawda M, Thakur K, et al. Physicochemical characterization of Suvarna Bhasma, its toxicity profiling in rat and behavioural assessment in zebrafish model. Journal of Ethnopharmacology. 2020;**249**:112388. DOI: 10.1016/j.jep.2019.112388

[13] Balkrishna A, Tomer M, Manik M, Srivastava J, Dev R, Haldar S, et al. Chyawanprash, an ancient Indian Ayurvedic medicinal food, regulates immune response in zebrafish model of inflammation by moderating inflammatory biomarkers. Frontiers in Pharmacology. 2021a;**12**. DOI: 10.3389/fphar.2021.751576

[14] Shanmugam R, Arthi B, Pragya P, Girigoswami A, Koyeli G, Pemula G, et al. A comprehensive assessment of the antioxidant capacity of varied Bacopa monnieri extracts and its toxicity on early

life stages of zebrafish embryos. Natural Product Communications. 2025;**20**(3):1. DOI: 10.1177/1934578X251327815

[15] Lakshmanagowda NK, Sagar N, Puttasiddaiah R, Sridhar K, Raghavendra VB, Bhaswant M. Benincasa hispida alleviates stress and anxiety in a zebrafish (Danio rerio) model. Life. 2024;**14**(3):379. DOI: 10.3390/life14030379

[16] Balkrishna A, Lochab S, Joshi M, Srivastava J, Varshney A. Divya-herbal-peya decoction harmonizes the inflammatory response in lipopolysaccharide-induced zebrafish model. Journal of Experimental Pharmacology. 2021b;**13**:937-955. DOI: 10.2147/JEP.S328864

[17] Wang M, Wang X, Li X, Dong Y, Yu J, Li P, et al. Chenopodium album L. herb attenuates inflammation by modulating multiple signaling pathways in zebrafish: Network pharmacology analysis, transcriptomics analysis and experimental verification. Journal of Functional Foods. 2025;**127**:106719. DOI: 10.1016/j.jff.2025.106719

[18] Kim S, Kim M, Kang M-C, Lee HHL, Cho CH, Choi I, et al. Antioxidant effects of turmeric leaf extract against hydrogen peroxide-induced oxidative stress in vitro in vero cells and in vivo in zebrafish. Antioxidants. 2021;**10**(1):112. DOI: 10.3390/antiox10010112

[19] de Ruyck J, Brysbaert G, Blossey R, Lensink MF. Molecular docking as a popular tool in drug design, an in silico travel. In: Advances and Applications in Bioinformatics and Chemistry. Vol. 9, No. 1. Macclesfield, Cheshire, United Kingdom: Dove Medical Press Ltd.; 2016. DOI: 10.2147/AABC.S105289

[20] Jakhar R, Dangi M, Khichi A, Chhillar AK. Relevance of molecular docking studies in drug designing. Current Bioinformatics. 2020;**15**(4):270-278. DOI: 10.2174/1574893615666191219094216

[21] Mat Zian NFA, Swain P, Mohd Faudzi SM, Zakaria N, Wan Ibrahim WN, Abu Bakar N, et al. Mapping molecular networks within Clitoria ternatea Linn. against LPS-induced neuroinflammation in microglial cells, with molecular docking and In vivo toxicity assessment in zebrafish. Pharmaceuticals. 2022;**15**(4):467. DOI: 10.3390/ph15040467

[22] Siddhu NSS, Guru A, Satish Kumar RC, Almutairi BO, Almutairi MH, Juliet A, et al. Pro-inflammatory cytokine molecules from Boswellia serrate suppresses lipopolysaccharides induced inflammation demonstrated in an in-vivo zebrafish larval model. Molecular Biology Reports. 2022;**49**(8):7425-7435. DOI: 10.1007/s11033-022-07544-5

[23] Bouabdallah S, Ibrahim MH, Brinza I, Boiangiu RS, Honceriu I, Amin A, et al. Anxiolytic and antidepressant effects of tribulus terrestris ethanolic extract in scopolamine-induced amnesia in zebrafish: supported by molecular docking investigation targeting monoamine oxidase A. Pharmaceuticals. 2024;**17**(9):1208. DOI: 10.3390/ph17091208

[24] Singh P, Hassan S, Kumar VR, Biswas K. General introduction to different neurodegenerative diseases. In: Neurodegenerative Diseases. CRC Press; 2025. pp. 1-19

[25] Galvin JE. Cognitive change in Parkinson disease. Alzheimer Disease & Associated Disorders. 2006;**20**(4):302-310. DOI: 10.1097/01.wad.0000213858.27731.f8

[26] Borah A, Choudhury A, Paul R, Mazumder MK, Chetia S. Neuroprotective effect of Ayurvedic preparations and natural products on Parkinson's disease. Neuroprotective Natural Products: Clinical Aspects and Mode of Action. 2017:91-105

[27] Biswas S, Davies LN, Sheppard AL, Logan NS, Wolffsohn JS. Utility of artificial intelligence-based large language models in ophthalmic care. Ophthalmic and Physiological Optics. 2024;**44**(3):641-671. DOI: 10.1111/opo.13284

[28] Chahardehi AM, Hosseini Y, Mahdavi SM, Naseh I. The zebrafish model as a new discovery path for medicinal plants in the treatment of Parkinson's disease. CNS & Neurological Disorders Drug Targets. 2024;**23**(3):306-314

[29] Verma B, Singh C, Singh A. Effect of hydro-alcoholic extract of Centella asiatica on streptozotocin induced memory dysfunction in adult zebrafish. Current Research in Behavioral Sciences. 2021;**2**:100059

[30] Lam CS, Korzh V, Strahle U. Zebrafish embryos are susceptible to the dopaminergic neurotoxin MPTP. European Journal of Neuroscience. 2005;**21**(6):1758-1762. DOI: 10.1111/j.1460-9568.2005.03988.x

[31] Dogra N, Nagpal D, Aeri V, Ahmad S, Pande Katare D. Evaluating the synergistic effect of Mucuna prurines extract and sesame oil against the Parkinson's disease zebrafish model: In-vivo/in-silico approach. All Life. 2021;**14**(1):1022-1042

[32] Carrasco-Gallardo C, Guzmán L, Maccioni RB. Shilajit: A natural phytocomplex with potential procognitive activity. International Journal of Alzheimer's Disease. 2012;**2012**(1):674142

[33] Katare DP. Protective effects of quercetin in zebrafish model of alzheimerâ€™ s disease. Asian Journal of Pharmaceutics (AJP). 2018;**12**(02)

[34] Prema S, Sharma S, Kondrapu P, Kumre YM, Salim MR, Khatkar PK, et al. Evaluation of antioxidant, antidiabetic and antihyperlipidemic activity of Syzygium cumini seeds in diabetic zebrafish model. European Chemical Bulletin. 2023;**12**(6):2833-2839

[35] Kumar A, Pawar N, Aole S, Faisal F, Nadkar V. Pharmacological investigation of antioxidants and antidiabetic activity of vitex trifolia bark in diabetic zebrafish model. Journal of Advanced Zoology. 2023;**44**(S-2):2370-2378

[36] Pérez Gutiérrez RM, Muñiz-Ramirez A, Garcia-Campoy AH, Mota Flores JM. Evaluation of the antidiabetic potential of extracts of Urtica dioica, Apium graveolens, and Zingiber officinale in mice, zebrafish, and pancreatic β-cell. Plants. 2021;**10**(7):1438

[37] Jin S, Hong JH, Jung SH, Cho KH. Turmeric and laurel aqueous extracts exhibit in vitro anti-atherosclerotic activity and in vivo hypolipidemic effects in a zebrafish model. Journal of Medicinal Food. 2011;**14**(3):247-256

[38] Pal M, Chaudhary P, Solanki SK. Ayurveda perspective on nidan, samprapti and chikitsa of poly cystic ovarian syndrome (PCOS). European Journal of Biomedical. 2020;7(11):258-262

[39] Balkrishna A, Nain P, Joshi M, Kumar B, Varshney A. Super-critical fluid extract of Bryonopsis laciniosa (Shivlingi) seeds restores fertility in zebrafish models through revival of

cytological and anatomical features. Journal of Ovarian Research. 2022;**15**(1):46

[40] Balkrishna A, Nain P, Joshi M, Khandrika L, Varshney A. Supercritical fluid extract of putranjiva roxburghii wall. Seeds mitigates fertility impairment in a zebrafish model. Molecules. 2021;**26**(4):1020

[41] Himaja K, Veerapandiyan K, Usha B. Aromatase inhibitors identified from Saraca asoca to treat infertility in women with polycystic ovary syndrome via in silico and in vivo studies. Journal of Biomolecular Structure and Dynamics. 2024:1-16

[42] Adeva-Andany MM, Pérez-Felpete N, Fernández-Fernández C, Donapetry-García C, Pazos-García C. Liver glucose metabolism in humans. Bioscience Reports. 2016;**36**(6). DOI: 10.1042/BSR20160385

[43] Qiu L, Jia K, Huang L, Liao X, Guo X, Lu H. Hepatotoxicity of tricyclazole in zebrafish (Danio rerio). Chemosphere. 2019;**232**:171-179. DOI: 10.1016/j.chemosphere.2019.05.159

[44] Sinha RA, Singh BK, Yen PM. Thyroid hormone regulation of hepatic lipid and carbohydrate metabolism. In: Trends in Endocrinology and Metabolism. Vol. 25, No. 10. Amsterdam, Netherlands: Elsevier Inc.; 2014. pp. 538-545. DOI: 10.1016/j.tem.2014.07.001

[45] Bartl M, Pfaff M, Ghallab A, Driesch D, Henkel SG, Hengstler JG, et al. Optimality in the zonation of ammonia detoxification in rodent liver. Archives of Toxicology. 2015;**89**(11):2069-2078. DOI: 10.1007/s00204-015-1596-4

[46] Bunchorntavakul C, Reddy KR. Review article: Herbal and dietary supplement hepatotoxicity. Alimentary Pharmacology & Therapeutics. 2013;**37**(1):3-17. DOI: 10.1111/apt.12109

[47] Teschke R, Eickhoff A, Wolff A, Frenzel C, Schulze J. Herbal hepatotoxicity and WHO global introspection method. Concise Review. 2013;**12**(1):11-21

[48] Brownlie A, Zon L. The zebrafish as a model system for the study of hematopoiesis. Bioscience. 1999;**49**(5):382-392. DOI: 10.2307/1313631

[49] Hill AJ, Heiden TCK, Heideman W, Peterson RE. Potential roles of Arnt2 in zebrafish larval development. Zebrafish. 2009;**6**(1):79-91. DOI: 10.1089/zeb.2008.0536

[50] Lee Y-J, Choi H-S, Seo M-J, Jeon H-J, Kim K-J, Lee B-Y. Kaempferol suppresses lipid accumulation by inhibiting early adipogenesis in 3T3-L1 cells and zebrafish. Food & Function. 2015;**6**(8):2824-2833. DOI: 10.1039/C5FO00481K

[51] Field HA, Ober EA, Roeser T, Stainier DYR. Formation of the digestive system in zebrafish I. liver morphogenesis. Developmental Biology. 2003;**253**(2):279-290. DOI: 10.1016/S0012-1606(02)00017-9

[52] Hinton DE, Couch JA. Architectural pattern, tissue and cellular morphology in livers of fishes: Relationship to experimentally-induced neoplastic responses. Fish Ecotoxicology. 1998;**86**:141-164

[53] Mesens N, Crawford AD, Menke A, Hung PD, Van Goethem F, Nuyts R, et al. Are zebrafish larvae suitable for assessing the hepatotoxicity potential of drug candidates? Journal of Applied Toxicology. 2015;**35**(9):1017-1029. DOI: 10.1002/jat.3091

[54] Vliegenthart ADB, Wei C, Buckley C, Berends C, de Potter CMJ, Schneemann S, et al. Characterization of triptolide-induced hepatotoxicity by imaging and Transcriptomics in a novel zebrafish model. Toxicological Sciences. 2017;**159**(2):380-391. DOI: 10.1093/toxsci/kfx144

[55] North TE, Babu IR, Vedder LM, Lord AM, Wishnok JS, Tannenbaum SR, et al. PGE2-regulated wnt signaling and N -acetylcysteine are synergistically hepatoprotective in zebrafish acetaminophen injury. Proceedings of the National Academy of Sciences. 2010;**107**(40):17315-17320. DOI: 10.1073/pnas.1008209107

[56] Zhang Y, Han L, He Q, Chen W, Sun C, Wang X, et al. A rapid assessment for predicting drug-induced hepatotoxicity using zebrafish. Journal of Pharmacological and Toxicological Methods. 2017;**84**:102-110. DOI: 10.1016/j.vascn.2016.12.002

[57] Zhang Y, Liu K, Hassan HM, Guo H, Ding P, Han L, et al. Liver fatty acid binding protein deficiency provokes oxidative stress, inflammation, and apoptosis-mediated hepatotoxicity induced by pyrazinamide in zebrafish larvae. Antimicrobial Agents and Chemotherapy. 2016;**60**(12):7347-7356. DOI: 10.1128/AAC.01693-16

[58] Balkrishna A, Bhattacharya K, Sinha S, Dev R, Srivastava J, Singh P, et al. Apparent hepatotoxicity of Giloy (Tinospora cordifolia): Far from what meets the eyes. Journal of Clinical and Experimental Hepatology. 2022;**12**(1):239

[59] Katoch S, Chhimwal J, Singh D, Kumar D, Patial V. Picrosides-rich fraction from Picrorhiza kurroa attenuates steatohepatitis in zebrafish and mice by modulating lipid metabolism and inflammation. Phytomedicine. 2025;**137**:156368. DOI: 10.1016/j.phymed.2025.156368

[60] Balkrishna A, Sharma N, Srivastava D, Kukreti A, Srivastava S, Arya V. Exploring the safety, efficacy, and bioactivity of herbal medicines: Bridging traditional wisdom and modern science in healthcare. Future Integrative Medicine. 2024;**3**(1):35-49. DOI: 10.14218/fim.2023.00086

[61] Yadav AG, Gandhi PK, Lagad CE, Ingole RK, Rajurkar SR. Comparison of Hepatoprotective activity of Dhatri Lauha (DL) and Dhatri Lauha without Lauha Bhasma (DLWLB) on carbon tetrachloride induced hepatotoxicity in wister rats. Journal of Ayurveda. 2024;**18**(4):288-292. DOI: 10.4103/joa.joa_237_23

[62] Honwad Scholar S, Shridhara Bairy T, Honwad S, Ravishankar B. Hepatoprotective activity of Somanathi Tamra Bhasma in paracetamol induced liver toxicity in albino-rats. The Journal of Phytopharmacology. 2015;**4**(3):143-146. Available from: www.phytopharmajournal.com

[63] Balkrishna A, Rustagi Y, Bhattacharya K, Varshney A. Application of zebrafish model in the suppression of drug-induced cardiac hypertrophy by traditional Indian medicine Yogendra Ras. Biomolecules. 2020;**10**(4):600

[64] Dinesh D, Nanjappa DP, Babu N, Kalladka K, Chakraborty G, Chakraborty A. Evaluation of toxicity and antiangiogenic activity of Murraya koenigii leaf extracts in zebrafish. Journal of Health and Allied Sciences NU. 2020;**10**(02):79-85

[65] Balkrishna A, Rustagi Y, Tomer M, Pokhrel S, Bhattacharya K, Varshney A. Divya-Arjuna-Kwath (Terminalia

arjuna) and Divya-HridyAmrit-Vati ameliorate isoproterenol-induced hypertrophy in murine cardiomyocytes through modulation of oxidative stress. Phytomedicine Plus. 2021;**1**(4):100074

[66] Riaz M, Khalid R, Afzal M, Anjum F, Fatima H, Zia S, et al. Phytobioactive compounds as therapeutic agents for human diseases: A review. Food Science & Nutrition. 2023;**11**(6):2500-2529. DOI: 10.1002/fsn3.3308

[67] Huang S, Beevers. Pharmacological and clinical properties of curcumin. Botanics: Targets and Therapy. 2011;**5**:5-18. DOI: 10.2147/BTAT.S17244

[68] Tang H, Qin N, Rao C, Zhu J, Wang H, Hu G. Screening of potential anti-thrombotic ingredients from salvia miltiorrhiza in zebrafish and by molecular docking. Molecules. 2021;**26**(22):6807. DOI: 10.3390/molecules26226807

Chapter 6

Health and Welfare in Early Stages of Zebrafish: Factors and Indicators

Juan Ramos, Ali Reza Khansari and Lluis Tort

Abstract

The early stages of zebrafish are very plastic in response to environmental and social conditions. Many factors can modify the normal development of zebrafish, including transgenerational factors related to their parents. Standardizing these conditions and controlling the environment must be done to ensure the healthy development of zebrafish. Handling during the developmental stages of zebrafish also alters the normal development and could interfere with their welfare. Traditional protocols like bleaching or anesthesia can have a deep impact on their welfare and could also affect future generations. To ensure a healthy colony, we must also consider how the colony and the broodstock will be managed in order to avoid endogamy and unhealthy embryos. There are many symptoms of these interferences in the normal development, such as morphological, behavioral, genetic, and hormonal alterations, that could be checked in the gametes or during the development. A standardization effort should be made to promote the reproducibility of experiments and the healthy development of zebrafish.

Keywords: health, welfare, early developmental stages, zebrafish, protocols, indicators

1. Introduction

A main part of zebrafish experiments associated with medical and biomedical research involves early developmental stages of zebrafish, so the ability to obtain healthy embryos will ensure the viability and reproducibility of zebrafish research [1]. Health indicators are an essential tool for evaluating developmental progress and detecting signs of stress or pathological problems. These indicators include a wide spectrum of morphological features and more refined measures such as behavior or genetic expression. Considering the factors that could influence the healthy development of zebrafish, they are not related just to the environmental conditions during embryonic development, because parental experience and gamete quality also determine the health and viability of the offspring. A proper standardization of husbandry protocols, such as environmental conditions, water, and feeding quality during parental and embryo life, is a necessary step for the zebrafish community [2]. Evaluating the quality of gametes and embryos will also be a functional tool in order to refine husbandry and prevent health problems in the future.

This chapter aims to go through an intergenerational approach and summarize the factors that allow for healthy improvement and standardized development. Parental influence, male and female (with special attention to nutritional state), water quality during development, experimental manipulation, and genetic conditions will be considered in this chapter. Establishing standardized criteria for assessing embryonic health is fundamental not only in order to assure reproducibility but also from an ethical point of view. Finally, we discuss how the health status of the gametes and embryos can be evaluated, and we propose a standard protocol to obtain healthy embryos.

2. Parental influence

Before egg fertilization, there are many influences on offspring viability. A holistic approach for health and welfare should be taken into account, as many transgenerational effects could compromise the healthy development of zebrafish embryos. Embryos could be influenced by their parents' previous experiences, modifying their normal development. There are different mechanisms for passing parental experiences to offspring. For example, the amount and the quality of the nutrients that gametes have, epigenetic changes in parents, or the number and quantity of hormones, such as cortisol and thyroid hormones [3]. So, for maintaining health in the early stages of zebrafish, we should consider parental health before generating the gametes. This standardization effort should be made already on the parental generation to set nutritional conditions, husbandry protocols, and age, or minimize stressors and pollutants for preserving the health status of the offspring. Also, this effort should be focused on maintaining the reproducibility of the experiences, as zebrafish have shown transgenerational plasticity.

2.1 Nutritional conditions

Zebrafish have external fecundation and development, so the nutrients for embryonic development till external feeding, which occurs at day 5 postfertilization, are provided by parents, especially by the females. Parental nutrition will determine reproductive performance and survival rates of the embryos. The nutritional state of the parents prior to spawning will be critical for the quality of the gametes, determining factors such as sperm motility or yolk sac content, which are used for gamete evaluation.

Maternal nutrition is crucial for oocyte quality and posterior embryogenesis. Zebrafish females that have been fed with high-quality and balanced diets produce eggs with better structural integrity, higher yolk volume, and more sperm capability, which will determine the fertilization success. On the other hand, if the diet is not well balanced, overfed females will have earlier ovary development but with a lower egg quality; the gametes obtained from unbalanced diets will have lower size and hatching rates [4].

Some micronutrients are crucial for the normal development and viability of zebrafish embryos. The lack of vitamin E will cause an increased peroxidation of the lipids, leading to a high mortality in the embryos. Selenium deficiency has been described as a source of increasing oxidative stress, causing a delayed development of embryos. Essential fatty acids, such as docosahexaenoic (DHA) and arachidonic acid, are well-known micronutrients in fish culture, as they are vital for fish embryonic development and should be provided in the food, as they cannot be synthesized by the fish themselves. Live food, and especially *Artemia*, is used in zebrafish facilities as environmental enrichment but also as a boost for the availability of fatty acids.

Artemia has been proven to increase reproduction rates and embryo survival, and it is regularly used in broodstock management [5].

Reproduction is the last physiological function along the lifespan of fish, so an optimal nutritional state is required for reproducing zebrafish. After a fasting period, especially for females, fish will employ the nutrients to restore the energy balance, and eggs obtained will have lower quality and reduced viability, so enough energy resources should be provided to them [6].

Paternal nutrition is also a critical factor for the offspring's development. Nutritional state affects sperm quality and motility, transmitting epigenetic information [7]. Studies have also described that paternal undernutrition, as in females, can lead to altered sperm, which affects energetic metabolism in the embryos. These changes will affect the growth and stress resilience of the offspring. These disorders, like metabolic diseases, can be transmitted to different generations and are heritable through epigenetic modifications [8]. So, epigenetic changes made by nutritional state have become an emerging area of interest [9, 10]. In addition, parental nutrition can modulate hormonal and metabolic pathways that influence gametogenesis and embryonic patterning. Nutritional imbalances may disrupt endocrine regulation, leading to impaired fertilization, altered sex ratios, or increased rates of embryonic malformation.

Parental pairs, under a balanced feeding regimen, lay eggs with higher survival rates and improved post-larval performance. Different feeding regimes, including commercial diets, have shown that they influence larval growth rates, pigmentation, behavior, and gene expression. Nowadays, different diets are available, and they will determine reproduction and embryo development, so a tested formula should be selected to ensure embryo health and consistent results [11].

2.2 Parental influence beyond nutrition on offspring health in zebrafish

Parental effects on offspring's health go much further than the nutritional state. In zebrafish, both paternal and maternal contributions will shape offspring phenotype, not just through a genetic pathway. Parents will influence the offspring with epigenetic inheritance and gamete quality (previously mentioned), but also in the stress physiology, immune status, and environmental exposures. So, the ability to tolerate infections and cope with new stimuli is related to parents' previous experience [12].

Usually, maternal effects are stronger than paternal ones, as in oviparous species, like zebrafish, females provide more genetic or nutritional material than males, including RNA, proteins, hormones, and mitochondria. During the first 3 hours of development, prior to the zygotic activation, the maternal molecules will play a central role till mid-blastula. The maternal transcription of genes such as *nanog*, *sox19b*, and *pou5f3* will initiate embryogenesis and cellular differentiation, and interference in their expression will lead to embryonic lethality and morphogenesis effects [13].

Maternal material will provide the bioenergetic capacity, as the number of mitochondria and their function regulate adenosine triphosphate (ATP) provision, calcium buffering, and apoptotic pathways. Any dysfunction in these mitochondria's capability, like increased oxidative damage, will affect germ cells. This damage could be induced by age, environmental pollutants, or chronic stress, which leads to an increased mortality or a delayed development in the early stages of zebrafish. Maternal age is a crucial factor that determines the fertility and viability of embryos, as the expression of mitochondria- and telomere-related genes (*tomm40*, *mpc2*, *nbn*) decreases their expression, affecting offspring viability [14].

Maternal influences are also related to cortisol deposition and thyroid hormones, which will play a crucial role in the embryogenesis of the offspring. Elevated concentrations of maternal cortisol, usually related to chronic stress, are transferred to the egg, interfering with the development of the hypothalamic-pituitary-internal (HPI) axis. Maternal cortisol concentration will influence the reactiveness of the offspring and the coping stress strategies. Cortisol level is also related to the immune function and will be affected by maternal experience and their stress level [15]. Also, if maternal thyroid function is altered, like in hypothyroidism, it will interfere with the normal development of somites and the nervous system. Maternal thyroid hormones have been demonstrated to be essential for embryonic development, and till they are secreted by the offspring, their levels and function are determined by maternal concentrations. The influence of paternal status has also been demonstrated, and as females, stressed males create offspring more anxious and with less resilience to stress [16].

Parental exposure to environmental pollutants, such as pharmaceutical drugs, microplastics, and heavy metals, causes intergenerational effects. Adult zebrafish that have been exposed to triclosan or bisphenol show reproductive impairment [17, 18]. The offspring generated will have developmental defects, reduced hatching rates, altered responses to stressful stimuli, and immune alterations. The transgenerational effects are passed to next generations by epigenetic reprogramming changes in germ cells, including histones and DNA methylation.

3. Influence of environmental conditions on the health status of early developmental stages

Early stages of zebrafish are very plastic as they try to cope with environmental conditions to adapt to them. Environmental and husbandry conditions can significantly influence the viability, morphology, behavior, gene expression, and physiology of the embryos at later stages of development. Controlling environmental conditions during incubation is essential not only for improving laboratory and reproducibility practices but also for preserving the health and welfare status of zebrafish.

3.1 Temperature

It is the most important abiotic factor in zebrafish embryogenesis, as larvae cannot regulate temperature, as they are ectothermic. If the temperature is not in the optimal range, it will have a deep impact on the embryo. The standardized temperature for embryos is 28.5°C, a temperature that allows embryonic development and reproducibility. If the temperature is above 30°C, the embryonic development will be faster, but it may induce cellular stress or malformations [19]. On the other hand, low temperatures, like 24°C, will slow the development, causing an impairment in the organogenesis.

Temperature has also been described as a critical factor for determining sex in zebrafish. High temperatures, over 35°C, during development, from 10 to 35 days, modify the normal sex ratio, promoting males [20]. These effects are transgenerational, and they last for at least one generation [21]. Sex determination is determined by multifactorial elements, but temperature is crucial and could be easily manipulated to promote an optimal ratio. Thermal stress can also modify gene expression of heat shock proteins or metabolic enzymes, modifying the normal physiology of developmental stages. Zebrafish embryos, as they try to cope with environmental conditions, will adapt their energy metabolism, angiogenesis, or stress response, for example, to adapt to new environmental conditions [22].

3.2 Water composition

Also plays an important role in preserving osmotic balance and allowing normal development of zebrafish embryos. There are many parameters, but the main ones are pH, conductivity, hardness, and concentrations of some essential ions like calcium, magnesium, and potassium. Water quality could disrupt the epithelial barrier, interfering with the ion transport function, which causes edema and chorion alterations. PH modifies oxygen concentration and enzyme function [23].

3.3 Oxygen viability

It is required by zebrafish embryos for metabolic function and overall viability. Hypoxia regulates gene expression, as the embryo tries to cope with the environmental conditions, inducing an upregulation of hypoxia inducible factor (*hifa, vegfa*) and adaptation for further hypoxic conditions. Hypoxia also causes retarded development and mortality when embryos are not able to cope with hypoxic conditions [24, 25].

3.4 Nitrogen compounds

Early stages of zebrafish, till day 16 postfertilization, breathe through the skin, so the gills are not so sensitive to nitrogen compounds as in the adult stage. Even though zebrafish embryos have shown induced oxidative stress and mitochondrial dysfunction in ionocytes when they are exposed to ammonia [26]. Nitrite from 400 mg/l causes craniofacial and axial malformations and alterations in the swim bladder formation and inflation, and similar effects have been observed with nitrates [27]. To prevent the negative effects of nitrogen compounds during zebrafish development, nitrogen compounds should be kept under 200 ppm.

3.5 Light

Controls the circadian rhythm of zebrafish, even during the early stages of development, as during the segmentation period, circadian genes are modulated by the photoperiod. These genes, *per1*, *clock*, and *bmal1*, regulate metabolic rate, cell cycle, and neurodevelopment. The absence of photoperiod disrupts the circadian rhythm, modifying behavioral phenotypes and causing malformations and unviable embryos [28]. Maintaining a natural photoperiod should be a priority, so egg incubators should have the capability of setting a 12:12 or 10:14 dark/light cycle. The type of light has also been proven to be important for the healthy development of zebrafish, and red light is not as compatible with normal development as white or blue light is. Also, UV light causes mortality and abnormalities such as pericardial edema and deformities [24].

3.6 Noise vibrations and mechanical stress

In zebrafish facilities, there are a lot of sources of noise and vibrations, such as pumps, aerators, water flow, or regular work such as tank removal and cleaning. Zebrafish become sensitive to vibrations and noise during early stages, as they develop the lateral line and neuromast, which start to develop at 18–24 h.

Exposure to vibration or noise during early developmental stages causes stress and alters normal development. Delayed hatching, axial deformity, pericardial edema, reduced activity, and increased mortality can be caused by noise and vibrations. Interference in the stress gene expression, an increased cortisol level, and an increased heart rate have also been noticed due to the noise and vibrations.

3.7 Environmental enrichment

The complexity of environmental stimuli, as they have in their natural biotope, enhances the proactive behavior and increases neural connections. Barrier environments, as they are traditionally set at the facilities, promote anxiety-like behavior [29]. Introducing different physical structures, such as substrates, during the incubation has been proven to enhance welfare in salmonids, but no specific protocol has been designed for the early stages of zebrafish [30].

3.8 Microbiological condition

The gut microbiota of adult zebrafish is composed mainly of proteobacteria, fusobacteria, and firmicutes. This microbiota can be distinct from one facility to another, depending on the diet, and is usually vertically transmitted from the broodstock to eggs. The gut microbial composition of parents also determines the gene expression of the offspring, particularly in genes related to the inflammatory and immune response. Differences due to the gut microbiota are observed in innate immune response, neutrophil, and macrophage development, and also in the integrity of the gut barrier. Different gut microbiota experiments have been tried in zebrafish, obtaining gnotobiotic fish and using probiotics to somehow control gut microbiota.

Some pathogens have the ability to infect the offspring before development, as they can colonize the gametes. In *Pseudoloma neurophilia*, it has been proven to have vertical transmission, infecting oocytes of females. During the first developmental stages, granulomas could be noticed in the embryo. Bleaching is not totally effective for this kind of transmission, as the pathogen is already in the embryo.

4. Influence of experimental protocols

4.1 Bleaching

Zebrafish embryos are highly sensitive to chemicals, and the sensitivity is dose-, time-, and stage-dependent. Zebrafish are especially sensitive during the first 2 h, as they include zygotic activation, gastrulation, and early organogenesis. Any manipulation during this period could have a deep impact on the healthy development of zebrafish [31].

A classic protocol regularly performed is the bleaching protocol, which consists of a chlorine bath for the eggs to disinfect and remove possible pathogens and other bacteria. Bleaching is commonly performed at 6 or 24 hours; however, embryos are particularly vulnerable at these stages, which can lead not only to surface disinfection but also to malformations or embryonic lethality. Bleaching is not a fully efficient disinfectant protocol, as some pathogens, especially those with vertical transmission (*Pseudoloma neurophila*), could survive chlorine exposure. Other disinfectant protocols, such as iodine or hydrogen peroxide, are used, but they could also be harmful to embryos. A nonimpact protocol for disinfecting eggs should be developed so that pathogens can be removed from zebrafish eggs but without damaging the embryos. Meanwhile, bleaching is done regularly nowadays at the facilities, somehow affecting the health and reproducibility of the research [32].

4.2 Transport

Usually, among zebrafish facilities, there exists a high rate of strain exchange, as there are many strains that are generated and shared among them. As there are not enough zebrafish stock centers or companies that handle high numbers of strains, regular transport is common between labs. Exchanging embryos is preferred over shipping adults, as it is cheaper, more fish can be exchanged, and in an easy way. Nevertheless, there is no standard protocol for shipping embryos, and usually, embryos are bleached and placed in a hermetic container such as a cellular flask or Falcon tube. The containers are placed in a Styrofoam box and shipped, with 6–8 h of development [33].

During transport, no control over environmental conditions is exercised, as it is carried out in closed containers. As we mentioned before, environmental conditions should be maintained in an optimal range to avoid modifications and health problems in embryos, including temperature, oxygen, and pH. In order to have healthy development, environmental parameters should optimally be kept stable, avoiding fluctuations. During the first stages of development, gastrulation or early segmentation, zebrafish are especially susceptible to vibrations, which could be produced during transport. Also, providing embryos with a regular photoperiod should be desirable to promote the healthy development of embryos. Unfortunately, there are no standard protocols that avoid or decrease the impact of transporting embryos. Nonetheless, after receiving transported embryos, they should be carefully watched, and malformed or unviable embryos should be discarded.

4.3 Microinjection

Microinjection is also perceived as a minimally invasive technique that allows gene editing and the creation of transgenic strains of zebrafish. Microinjection is done in the early stages of development (first hour of development) and requires egg manipulation and placing them in a microscope slide, almost in dry conditions. Injecting the genetic material could have a deep impact on the embryo, depending on where it is injected (yolk or blastocyst) and the volume and composition of the material injected. Mechanical damage could be caused due to the injection, causing a disruption of the cell architecture, which could cause an arrest of cell signaling or alter apoptosis signaling. The impact of the microinjection could be transient or long-term, so after microinjection, control over embryo development should be performed. Also, the new strain generated, due to the inclusion of the genetic material, may experience physiological or behavioral modifications, so the new strain should be phenotyped.

4.4 Anesthesia and euthanasia

During the early stages of development, zebrafish are used for research, and imaging is a useful and regular technique, as embryos are translucent, so the development can be monitored in real time. For these imaging techniques, anesthesia is commonly used, as it allows immobilization of embryos. Unfortunately, anesthetizing embryos also has an impact on the health and welfare of zebrafish developmental stages, causing multiple modifications to the normal development. These modifications will be dependent on the anesthetic, the dose, the time of exposure, and the stage of development. The most commonly used anesthetics employed are tricaine, clove oil, lidocaine, and benzocaine, which, in high concentrations, could also be used as euthanasics. These compounds act by blocking sodium channels or gamma-Aminobutyric acid (GABA) receptors, which cause different modifications of normal development [34].

Embryos under anesthesia can develop different malformations, such as cardiac or yolk edema, craniofacial malformations or axis ones, delayed hatching, or alterations in neuronal development, for example. These malformations are accompanied by an altered gene expression in different key genes, which will interfere with the normal development, causing malformations, as anesthetics interfere with apoptotic pathways, cell differentiation, and organogenesis [35].

During the early stages of development, zebrafish are able to perceive noxious stimuli, escape from them (some chemicals are aversive to them), and process them as stressful stimuli, even before the HPI axis is well developed. Cortisol will be released by embryos from day 5, but before it, some early indicators could be used as they are already functional, such as *hsp* or *c-fos* genes. Specific anesthetic protocols for each developmental stage should be developed to avoid or minimize the impact on the health of embryos and also anesthetics, as they are able to process noxious stimuli.

4.5 Incubation

Incubation is not a proper experimental protocol, but as it requires handling, some considerations should be taken to keep embryos healthy. Controlled conditions should be established as soon as possible, and removing eggs from the tank should be a priority. Good breeders usually lay the eggs in 20 minutes, so 45 minutes will be enough time for parents to lay fertilized eggs.

Eggs have a negative buoyancy, so after being laid, they go to the bottom of the tank, with all the debris. Standard crossing tanks are static, so the water quality is not controlled, and nitrogen compounds increase their concentration, and oxygen and temperature decrease.

Removing the debris associated with the eggs is quite easy and will avoid the proliferation of fungus, ciliates, and free nematodes. Just pouring some clean fish water into the strainer with the eggs will break and remove the debris. The room and the water should be previously acclimated and conditioned to prevent sudden changes of temperature and water quality. Traditional embryo media is made artificial, and methylene blue is added to avoid fungal proliferation. Methylene blue has been proven to be mutagenic, so it should not be used routinely, and removing debris is a useful and non-damaging routine to avoid fungus proliferation.

Eggs are traditionally incubated in petri dishes in acclimated and photoperiod-controlled areas. Avoiding crowding in the petri dish, with too many eggs, will facilitate the daily care of the eggs and allow for good quality. Usually, about 50 eggs placed into a 9 cm petri dish show no interference in the healthy development of the eggs. Every day, the petri dish should be checked to remove dead embryos, chorions, and debris with a plastic Pasteur pipette, avoiding fungus proliferation and the degradation of the water quality. Also, 1/3 of the total volume should be renewed with clean water, decreasing nitrogen concentrations and the total live charge in the petri dish.

Nowadays, there is no validated and standardized protocol for incubation, and there are not enough scientific papers to validate one. The recommendations that are given in this section come from our personal experience and from unpublished data.

4.6 Emergency phenotype

Zebrafish have a huge plasticity, especially during the early stages of development. They try to cope with environmental conditions, modifying their genetic and physiological functions, but sometimes these conditions are too severe for embryos, so they hatch earlier in an attempt to escape from those conditions [12]. This plasticity has been described in salmonids as an emergency phenotype, as early stages of fish are free and can escape from those adverse conditions. In the case of zebrafish, we have some examples of this emergency phenotype; if they feel in danger of being predated, mechanical stress, or ionic stress, for example, they will hatch without developing their swimming ability in an attempt to escape from that source of danger. So, an early hatch of zebrafish embryos should be interpreted as an unfavorable indicator of health and used to evaluate environmental or handling conditions [36].

5. Influence of genetic conditions

In zebrafish husbandry, keeping life strains is quite hard, as zebrafish do not tolerate inbreeding, and individual identification is not a practical issue, as in mice management. Inbreeding depression is translated as a loss of fitness population, as homozygosis and deleterious genes increase. Signs of inbreeding depression include gender biases (usually more males), a decrease in fertility, growth, and survival, and more malformations. Inbreeding morphological abnormalities are evident after six generations and could be more pronounced as the fish get older. Skeletal malformations, such as spinal curvatures, caudal fin developmental deformities, craniofacial malformations, especially in the jaws, and improper development of the opercula, are usually caused by inbreeding [37].

Genetic variation can be established by creating hybrid lines; however, this could be a source for outbreeding depression by crossing distantly related lines. If these lines are not related, a heterozygosity line will be obtained, and the fitness of the strain will be increased by a phenomenon named hybrid vigor, where the offspring are superior to the parents. This hybridization can also lead to a loss of locally adapted alleles.

For line maintenance, there is no magic formula, but selecting breeders and offspring and avoiding the malformations and low-fitness parents should be a must. The first thing to do is to define how many breeders compose the brood stock and the method of crossing (single cross tank, mass crossing tank), so the number of crossing units to establish each generation could be defined. If the next generation is going to be obtained by a lot of crossing units, more genetic variability will be obtained, but always avoid using siblings.

In the zebrafish research community, there are no available companies, as in mice, that provide strains that are well-phenotyped and well-stabilized. Asking for new blood to refresh the inbreeding of the strains is a way of mitigating this genetic depression and could be obtained from other labs or a few genetic stock centers. Besides this problem, there is a huge number of strains that have been generated in zebrafish, which multiplies the effort that should be made to keep genetic quality. Generating transgenic strains, due to the CRISPR/Cas9 technique and the external development, is easy and cheap. Usually, these new strains are not phenotyped, so the implications in gametogenesis and embryogenesis are unknown. Phenotyping studies are multidisciplinary studies that should be done after generating a strain, but in zebrafish, they are not standardized [37].

6. Evaluation of gamete and embryo quality of zebrafish

As it has been described before, there are multiple factors that can influence the normal development of zebrafish, maintaining a good health status. Zebrafish are coping with environmental conditions and adapting their behavior, physiology, and morphology in a plastic manner, especially during gametogenesis, embryogenesis, and organogenesis. These adaptations could be physiologically or pathologically translated and expressed in different moments and manners. In this section, we summarize some indicators that are useful for zebrafish husbandry. Early evaluation of gametes and embryos should be implemented in standard husbandry to promote healthy zebrafish. To help with husbandry routines, a table summarizing all the indicators has been made as a practical tool for technicians (see **Table 1**).

Indicator	Description	Measurement
Sperm quality	Motility, morphology, and DNA integrity; affected by environmental stressors.	Microscopy (e.g., Makler chamber), Computer-Assisted Sperm Analysis (CASA), and DNA integrity assays.
Oocyte quality	Morphology and proteomic markers; impacted by nutrition and stress.	Morphological evaluation, proteomic analysis, and gene expression profiling.
Embryonic morphology		
• Chorion and perivitelline space	Abnormalities may indicate developmental issues due to calcium deficiency.	Light microscopy to observe structural integrity and spacing.
• Hatchability	Reduced or premature hatching can signal environmental stress or contamination.	Monitoring hatching rates and timing under controlled conditions.
• Epiboly and blastoderm morphology	Delays or asymmetries reflect suboptimal conditions affecting gastrulation.	Microscopic observation of cell movement and layer formation.
• Developmental delays	Deviations from standard timelines may indicate genetic defects or toxic exposures.	Time-lapse imaging to compare developmental milestones against standard timelines.
• Somite formation	Abnormalities suggest segmentation or axis formation issues.	Microscopy to count and assess somite structure and symmetry.
• Cardiac morphology and function	Malformations or edema indicate circulatory or developmental problems.	High-resolution imaging, heart rate measurement tools, and possibly automated systems.
• Edema	Pericardial and yolk sac edema are signs of physiological imbalance.	Visual inspection under a microscope for fluid accumulation.
• Yolk utilization	Delayed absorption or abnormalities suggest impaired nutrient utilization.	Observation of yolk absorption rates and morphology.

Indicator	Description	Measurement
• Spinal deformities and body curvature	Deformities indicate structural or neuromuscular issues.	Microscopic examination for structural abnormalities.
• Pigmentation defects	Irregular patterns may result from stress or genetic factors.	Visual assessment of pigmentation patterns and intensity.
Behavioral indicators		
• Spontaneous tail coiling (17–28 hpf)	Reflects spinal cord and neuromuscular integrity; reduced frequency may indicate toxicity.	Observation of tail movements under a microscope.
• Touch-evoked escape response (≥24 hpf)	Assesses muscle performance; impaired response may indicate developmental abnormalities.	Gentle tactile stimulation and response observation.
• Locomotion and startle response (≥3–5 dpf)	Altered activity can reflect exposure to neuroactive agents or stress.	Behavioral assays using video tracking systems.
• Swimming patterns (4–5 dpf)	Hypoactivity or erratic movement can indicate distress or developmental issues.	Automated tracking of swimming behavior using systems like ZebraBox.
Biochemical and genetic markers		
• Cortisol levels	Elevated levels indicate acute or chronic stress.	Whole-body cortisol extraction and quantification assays.
• Heat shock proteins (e.g., hsp70, hsp90)	Upregulated under stress, serving as indicators of cellular damage.	Gene expression analysis *via* qPCR or Western blotting.
• Apoptosis-related genes (e.g., bax, casp3a)	Increased expression signals cellular distress.	Gene expression profiling using qPCR.
• Inflammatory markers (e.g., il1b, tnfa)	Upregulation occurs following infection or chemical stress.	Measurement of gene expression levels through qPCR.
• Oxidative stress markers (e.g., sod1, cat)	Reflect the response to oxidative damage from pollutants or UV exposure.	Enzymatic activity assays and gene expression analysis.
• miRNA expression (e.g., miR-29a)	Overexpression is linked to stress and developmental malformations.	Quantitative polymerase chain reaction (PCR) for specific microRNA levels.

Table 1.
Indicators of stress and welfare in zebrafish early stages.

6.1 Gamete quality indicators

Sperm quality: environmental conditions could strongly compromise the viability of sperm and the offspring [9].

6.1.1 Classic evaluation

Usually done with a Makler chamber with a contrast microscope. The traditional parameters checked are concentration, volume (>3 microliter), and kinetics (>70% motility and time of activation). Also, malformations such as macro- or microcephaly, degenerated head, and broken or folded tail could be detected.

6.1.2 New parameter

Using software like Computer-Assisted Sperm Analysis (CASA) [7], the classic kinetics parameters could be better interpreted, obtaining values like curvilinear velocity, straight line velocity, and their average at different time points (15, 30, 60 s). DNA integrity could be evaluated with different methods related to the enzymatic targeting of DNA or chromatin susceptibility to DNA denaturation, and genetic information will be transferred to the next generation [38].

6.2 Oocyte quality

As sperm, many factors alter the viability and the quality of the oocyte. Nutrition is more essential in females than in males, as embryonic nutrition will depend exclusively on the yolk provision, and the feeding rate of reproductive females should be 5% of their weight per day.

6.2.1 Viability test

Trypan blue is used to evaluate the integrity of the membranes. Healthy and viable oocytes do not take the dye. Other dyes like fluorescein acetate or propidium iodide could be used to evaluate the integrity of the oocyte [39].

6.2.2 Developmental and fertilization rates

The number of fertilized eggs and their normal development could be assessed in order to control oocyte quality, but as it has been previously described, it also depends on sperm quality. A historical dataset of these rates to control oocyte quality could be of help, so the records could be compared.

6.2.3 Proteomics

Different proteins could be used as biomarkers of oocyte quality. These markers could be used for evaluating the maturation process of the oocyte.

6.3 Morphological evaluations

Morphology is one of the most direct and accessible methods for assessing embryonic and larval health. It can be done by classical microscopy or by automated systems that detect morphological features or by quantitative measurements. Some indicators have been developed for drug discovery and toxicity assays, such as the *Fish Embryo Toxicity Test* (FET) [40], but could be used for selecting healthy embryos [41].

6.3.1 Chorion and perivitelline space

The chorion is a hard protective membrane that helps the embryo during the first hours to isolate from the media. Some modifications in chorion hardness could interfere with the normal development, usually caused by a lack of calcium salts. During the first hours of development, a perivitelline space is formed between the chorion and the embryo. This space could collapse due to a toxicant or an osmotic imbalance [42].

6.3.2 Hatchability

A reduction in hatch rate or premature hatching may be indicative of environmental stress, microbial contamination, or impaired chorion function [43]. Zebrafish can hatch early as an adaptive response, which helps them to escape from predators or other dangers. Sometimes embryos cannot hatch on time, as bleaching or transport delays it, but an enzymatic bath of pronase could be done to remove the chorion artificially. The modification of the hatching rates and time of hatch has been used as a health indicator showing correspondence with further viability [36].

6.3.3 Epiboly and blastoderm morphology

Gastrulation is a particularly sensitive developmental window in zebrafish. Delays or asymmetries in epiboly are often used as indicators of healthy development. Morphogenetic movements during epiboly and convergence-extension are vulnerable to environmental conditions such as temperature fluctuations, hypoxia, or exposure to anesthetics. Delayed or asymmetric epiboly progression often correlates with reduced hatch rates and increased malformations later in development [44].

6.3.4 Developmental delays

Standard developmental conditions of zebrafish embryos were established by Kimmel, and it is used as a reference for embryo development. Deviations from the expected timing of events such as epiboly, somite formation, or hatching may reflect suboptimal incubation conditions, genetic defects, or toxic exposures [45, 46].

6.3.5 Somite formation

It begins after 10 h of fertilization, forming somites bilaterally along the axis. Changes in shape, bilaterally, or number indicate an impairment in segmentation or axis formation. The number of somites could be used as an indicator of development during embryogenesis [40].

6.3.6 Cardiac morphology and function

At 24 h of development, the heart becomes visible. Different malformations in the cardiac tube or pericardial edema can be noticed under basic optical microscopy. The development of some tools has helped to control heart rate in zebrafish embryos, such as confocal microscopy, echocardiography, and kymography [47].

6.3.7 Edema

Pericardial and yolk sac edema are common signs of physiological imbalance, often induced by chemical exposure or osmotic stress [48]. It may be caused by kidney or circulatory failure, leading to ionic imbalance and permeability changes.

6.3.8 Yolk

Energy provision under embryonic development depends exclusively on an efficient absorption of yolk nutrients. A delayed abortion, enlarged yolk, or detachment are indicators of impairment and delayed development. Yolk is also related to morphogenesis, so its normal development is necessary for a healthy embryo [41].

6.3.9 Spinal deformities and body curvature

These deformities can be induced by various factors such as lead exposure, genetic mutations, and disruptions in cerebrospinal fluid. Abnormalities such as axial curvature or scoliosis indicate structural or neuromuscular issues and could be generated by nutrient deficiencies, apoptosis, inflammatory processes, disruption of gene expression, or alterations in the central system fluid flow and cilia function (CSF) (see **Figure 1**) [40, 49].

6.3.10 Pigmentation defects

Could be an increased pigmentation (hyperpigmentation) or a reduced pigmentation (hypopigmentation). The pigmentation pattern is specific to the strain and could be really different from one line to another, such as *casper* or *leopard* lines. Irregular melanophore patterns or hypopigmentation may be associated with stress, mutations, or disrupted endocrine signaling. Altered melanophore distribution or pigmentation loss is associated with oxidative stress and altered neuroendocrine signaling [50].

6.4 Behavioral indicators

Despite their limited behavioral repertoire, zebrafish embryos and larvae display several measurable behaviors that can serve as indicators of health:

6.4.1 Spontaneous tail coiling (17: 28 hpf)

It is a simple movement originating at the spinal cord. This earliest form of motor activity reflects the integrity of the spinal cord and neuromuscular system. Reduced frequency or irregular patterns may signal developmental toxicity [51].

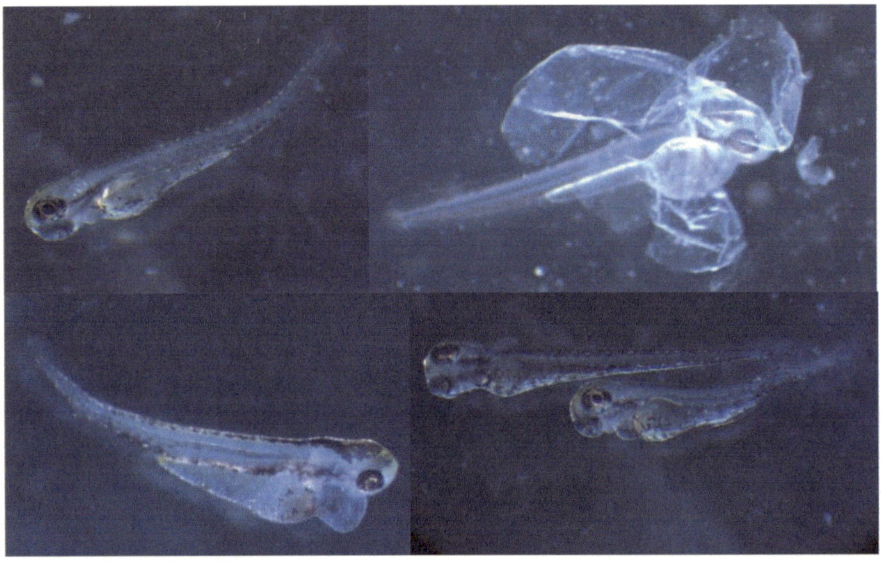

Figure 1.
*Different deformities found in zebrafish developmental stages: (**A**) Deformed axis, (**B**) hatching embryo, (**C**) severe cardiac edema and impaired digestive system, and (**D**) normal embryo and axis in "S" shape.*

6.4.2 Touch-evoked escape response (≥24 hpf)

Can be used to assess muscle performance during short movements [52]. This response to tactile stimuli can be impaired by sedation, neurotoxicity, or developmental abnormalities [52].

6.4.3 Locomotion and startle response (≥3: 5 dpf)

It is a rapid response triggered by various sensory stimuli. It may be a defense mechanism by which embryos could habituate under repeated exposure. It can be measured by quantifying the C-shaped body bend. Altered swimming activity, hypoactivity, or hyperactivity can reflect exposure to neuroactive agents, metabolic stress, or cardiovascular dysfunction [49].

6.4.4 Swimming patterns (4: 5 dpf)

To assess muscle performance during sustained periods. The development of swimming patterns is linked to the proper development of the nervous system and muscle coordination. The combination of different patterns, such as burst swimming, slow routine swimming (exploration pattern), and escape response in time and speed, could be used as a health indicator and neuromuscular activity and development. Hypoactivity or erratic locomotion can be signs of distress or exposure to toxic agents. Quantitative tracking systems have been developed to evaluate these parameters under controlled lighting and acoustic conditions [53].

6.5 Biochemical and genetic markers

6.5.1 Cortisol levels

It has been classically the main indicator for evaluating stress. In early stages, cortisol is not synthesized till day 5, and before, its concentration depends on maternal material. The concentration of maternal cortisol will influence the development of the copying style of the fish, for example, determining their anxiety or boldness. Till external feeding, the HPI axis and cortisol release will not be fully developed, so early stress cannot be evaluated by cortisol [15].

6.5.2 Heat shock protein expression (e.g., hsp70, hsp90)

In early stages, heat shock proteins play a crucial role in development, in cell movements and differentiation, and also as a cellular stress response. Classically, hsp has been used as a general indicator of cellular damage. Hsp genes are rapidly upregulated under stress and have been validated as general indicators of cell damage [54].

6.5.3 Apoptosis-related genes (e.g., bax, casp3a, casp9)

During embryogenesis, apoptotic genes play a crucial role in the normal development and morphogenesis. Apoptotic pathways are also related to stress, as they could be activated due to hypoxia, energetic impairment, and other environmental conditions. Stress could induce cell death in an intrinsic way (impairing mitochondrial physiology) or by activating specific cell death receptors in the extrinsic pathway [55].

6.5.4 Inflammatory markers (e.g., il1b, tnfa)

Stress, and particularly chronic stress, can lead to the activation of the pro-inflammatory cytokines, upregulating the inflammatory response at different levels. There are some cytokines that have a protective function, such as *IL-10*, protecting neurons from degeneration [56].

6.5.5 C-fos

As the HPI axis and central nervous system are not well developed till day 5, genes such as *c-fos* could be used as an early marker of neural activity, including neurons and glial cells. C-fos expression is currently used as a marker of neuronal activity linked with stress and anxiety. By monitoring *c-fos* expression, specific regions could be identified as they are activated due to stimuli such as a new object or a stressful stimulus.

7. A step toward standardization

Summarizing this chapter, as an effort to standardize husbandry conditions for the early developmental stages of zebrafish, we propose a guide for optimizing offspring welfare and health. Final protocols should be adapted to each facility and the resources available.

Obtaining eggs (see **Figure 2**):

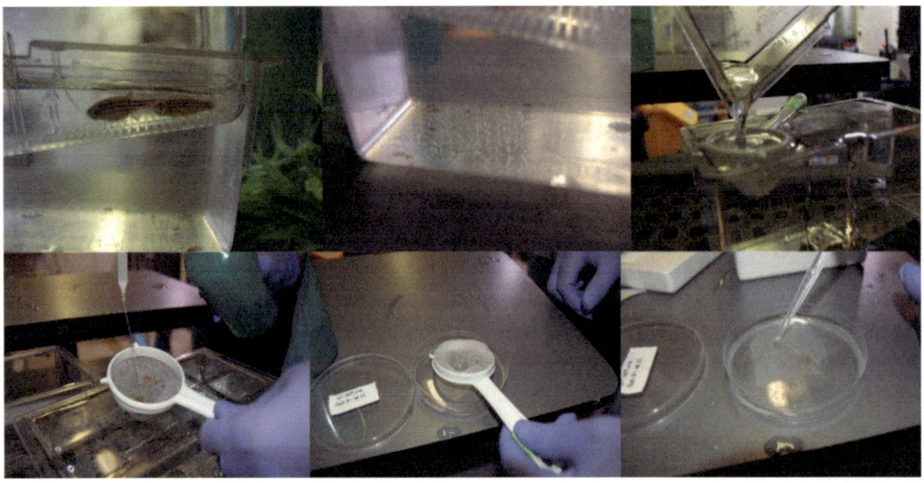

Figure 2.
Method for obtaining clean and healthy eggs. (**A**) *Crossing tank with parental pair and egg at the bottom,* (**B**) *egg and debris with negative buoyancy in the crossing tank,* (**C**) *egg transfer from crossing tank to egg strainer (wet to wet),* (**D**) *debris removal using a squeeze bottle,* (**E**) *egg transfer from strainer to petri dish, and* (**F**) *pipette cleaning: debris and unviable eggs.*

1. Parental husbandry
    a. High-quality food: dry and live (specific for zebrafish and developmental stage)
        i. Live food: Artemia at least 3 weeks before
        ii. Dry food: Five percent of body weight. At least two times
    b. Parental condition: Avoid stress and sibling mating.
2. Mating
    a. Use healthy and good-condition parents
    b. Fill the crossing tank with clean, acclimated, and UV-disinfected water
    c. Gentle fishing of parents
    d. No more than two fish per liter: O_2 depletion, nitrogen compounds, stress
    e. Place the crossing tank in a quiet, acclimated (26–28.5°C) specific area in the evening
    f. Crossing area photoperiod controlled, with a dark night
3. Collecting eggs
    a. Gently remove parents, avoid resuspending the eggs.
    b. Remove the egg from the tank as soon as possible.
    c. Use a semiflexible strainer, avoid eggs in dry conditions, to recollect the eggs from the bottom of the tank.
    d. Remove all the debris from the eggs, using a squeeze bottle filled with conditioned, acclimated, and clean water.
    e. Place some clean water in a clean petri dish.
    f. Turn the semiflexible strainer over the petri dish and trap the eggs with the superficial tension of the petri dish water.
    g. In a 9 cm diameter petri dish, place 50 healthy eggs. No debris inside.
    h. Fill the petri dish till ¾ parts, so air exchange can be done.
4. Bleaching
    a. Must only be done if it is necessary for barrier control reasons.
    b. It should be performed in chlorinated eggs of 24–30 h.
    c. Fill the container with acclimated solutions.

i. Container 1: fish water

ii. Container 2: fish water +30 ppm of chlorine (food-quality bleach).

iii. Container 3: fish water +60 ppm of sodium thiosulfate.

iv. Container 4: fish water

e. Remove unviable eggs from each petri dish, and place them into a mesh basket placed in container 1.

f. Place the mesh basket into container 2, with the bleach solution, for 10 minutes.

g. Transfer the basket into container 3 for 10 seconds.

h. Transfer the basket into container 4.

i. Transfer the eggs into a new petri dish with clean, acclimated water.

5. Incubation

a. Place the petri dish in a temperature-controlled room or an incubator (26–28.5°C) with a controlled photoperiod.

b. Every day, remove debris, chorion, and unviable embryos with a Pasteur pipette.

c. Renew 1/3 of the total water volume of the petri dish with new, clean water.

d. On day 5, embryos are able to eat independently, so they should be placed in the recirculation system. Acclimation for new water should be done by placing the petri dish floating on the tank surface and changing the embryo water with the system water gradually. After acclimation, larvae are placed in the tank by gently sinking the petri dish in the tank, which has a specific screen to avoid larva escape.

8. Conclusion

The early stages of zebrafish development are not covered by law requirements, so there are no legal restrictions till day 5 postfertilization. They are very plastic to environmental and parental conditions, hence adapting their genetic expression and physiology to cope with the environment. Many factors affect their normal development, such as environment, nutrition, parental state, and the handling procedures performed during early stages. Some of these conditions could impact their health and welfare, so an evaluation of the physiological stage should be performed, as they are a valuable tool to promote zebrafish welfare. However, such husbandry protocols are not yet standardized. As there are many ways of evaluating the quality and health of zebrafish embryos and gametes, standardization of husbandry protocols for early stages should be done by the zebrafish research community under scientific validation. This standardization will promote zebrafish welfare and reproducibility of the studies, pushing zebrafish to the next step as a biomedical model.

Author details

Juan Ramos, Ali Reza Khansari and Lluis Tort*
Department of Cell Biology, Physiology and Immunology, Universitat Autonoma de Barcelona, Spain

*Address all correspondence to: lluis.tort@uab.cat

References

[1] Choi TY, Choi TI, Lee YR, Choe SK, Kim CH. Zebrafish as an animal model for biomedical research. Experimental & Molecular Medicine. 2021;**53**(3):310-317. DOI: 10.1038/s12276-021-00571-5

[2] Tsang B, Zahid H, Ansari R, Lee RCY, Partap A, Gerlai R. Breeding zebrafish: A review of different methods and a discussion on standardization. Zebrafish. 2017;**14**(6):561-573. DOI: 10.1089/zeb.2017.1477

[3] Massey MD, Dalziel AC. Parental early life environments drive transgenerational plasticity of offspring metabolism in a freshwater fish (*Danio rerio*). Biology Letters. 2023;**19**(10):20230266. DOI: 10.1098/rsbl.2023.0266

[4] Dosch R. Next generation mothers: Maternal control of germline development in zebrafish. Critical Reviews in Biochemistry and Molecular Biology. 2015;**50**(1):54-68. DOI: 10.3109/10409238.2014.985816

[5] Nowosad J, Kucharczyk D, Targońska K. Enrichment of zebrafish Danio rerio (Hamilton, 1822) diet with polyunsaturated fatty acids improves fecundity and larvae quality. Zebrafish. 2017;**14**(4):364-370. DOI: 10.1089/zeb.2017.1416

[6] Reading BJ, Andersen LK, Ryu YW, Mushirobira Y, Todo T, Hiramatsu N. Oogenesis and egg quality in finfish: Yolk formation and other factors influencing female fertility. Fishes. 2018;**3**(4):45. DOI: 10.3390/fishes3040045

[7] Gallego V, Asturiano JF. Sperm motility in fish: Technical applications and perspectives through CASA-mot systems. Reproduction, Fertility and Development. 2018;**30**(6):820-832. DOI: 10.1071/RD17460

[8] Carvan MJ et al. Mercury-induced epigenetic transgenerational inheritance of abnormal neurobehavior is correlated with sperm epimutations in zebrafish. PLoS One. 2017;**12**(5):e0176155. DOI: 10.1371/journal.pone.0176155

[9] Riesco MF, Valcarce DG, Martínez-Vázquez JM, Robles V. Effect of low sperm quality on progeny: A study on zebrafish as model species. Scientific Reports. 2019;**9**(1):11192. DOI: 10.1038/s41598-019-47702-7

[10] Soubry A, Hoyo C, Jirtle RL, Murphy SK. A paternal environmental legacy: Evidence for epigenetic inheritance through the male germ line. BioEssays. 2014;**36**(4):359-371. DOI: 10.1002/bies.201300113

[11] Markovich ML, Rizzuto NV, Brown PB. Diet affects spawning in zebrafish. Zebrafish. 2007;**4**(1):69-74. DOI: 10.1089/zeb.2006.9993

[12] Ramos J, Balasch JC, Tort L. About welfare and stress in the early stages of fish. Frontiers in Veterinary Science. 2021;**8**:634434. DOI: 10.3389/fvets.2021.634434

[13] Pálfy M, Schulze G, Valen E, Vastenhouw NL. Chromatin accessibility established by Pou5f3, Sox19b and Nanog primes genes for activity during zebrafish genome activation. PLoS Genetics. 2020;**16**(1):e1008546. DOI: 10.1371/journal.pgen.1008546

[14] Zhu L, Lin Z, Liu Y, Sun H, Sun C, Chen F. Mechanisms of the effect of maternal age-related oocyte aging on fertility: Transcriptomic sequencing

analysis of a zebrafish model. Sichuan Da Xue Xue Bao Yi Xue Ban. 2024;**55**(3):588-595. DOI: 10.12182/20240560205

[15] Van Den Bos R et al. Early life exposure to cortisol in zebrafish (Danio rerio): Similarities and differences in behaviour and physiology between larvae of the AB and TL strains. Behavioural Pharmacology. 2019;**30**(2 and 3):260-271. DOI: 10.1097/FBP.0000000000000470

[16] Yeramilli V et al. Parental preconception stress in zebrafish induces long-lasting anxiety in offspring. Physiology & Behavior. 2024;**277**:114477. DOI: 10.1016/j.physbeh.2024.114477

[17] Ding P et al. Transgenerational thyroid hormone disruption in zebrafish induced by environmentally relevant concentrations of triclosan. Environmental Pollution. 2025;**374**:126242. DOI: 10.1016/j.envpol.2025.126242

[18] Mostari MH, Rahaman MM, Akhter MA, Ali MH, Sasanami T, Tokumoto T. Transgenerational effects of bisphenol A on zebrafish reproductive tissues and sperm motility. Reproductive Toxicology. 2022;**109**:31-38. DOI: 10.1016/j.reprotox.2022.02.005

[19] Pype C et al. Incubation at 32.5°C and above causes malformations in the zebrafish embryo. Reproductive Toxicology. 2015;**56**:56-63. DOI: 10.1016/j.reprotox.2015.05.006

[20] Ribas L, Liew WC, Díaz N, Sreenivasan R, Orbán L, Piferrer F. Heat-induced masculinization in domesticated zebrafish is family-specific & yields a set of different gonadal transcriptomes. Proceedings of the National Academy of Sciences of the United States of America. 2017;**114**(6):E941-E950. DOI: 10.1073/pnas.1609411114

[21] Valdivieso A, Ribas L, Monleón-Getino A, Orbán L, Piferrer F. Exposure of zebrafish to elevated temperature induces sex ratio shifts and alterations in the testicular epigenome of unexposed offspring. Environmental Research. 2020;**186**:109601. DOI: 10.1016/j.envres.2020.109601

[22] Schnurr ME, Yin Y, Scott GR. Temperature during embryonic development has persistent effects on metabolic enzymes in the muscle of zebrafish. Journal of Experimental Biology. 2014;**217**(8):1370-1380. DOI: 10.1242/jeb.094037

[23] Hammer HS. Water quality for zebrafish culture. In: The Zebrafish in Biomedical Research: Biology, Husbandry, Diseases, and Research Applications. San Diego, CA, USA: Elsevier; 2019. pp. 321-335. DOI: 10.1016/B978-0-12-812431-4.00029-4

[24] Andrade TS, Henriques JF, Almeida AR, Soares AMVM, Scholz S, Domingues I. Zebrafish embryo tolerance to environmental stress factors—Concentration–dose response analysis of oxygen limitation, pH, and UV-light irradiation. Environmental Toxicology and Chemistry. 2017;**36**(3):682-690. DOI: 10.1002/etc.3579

[25] Barrionuevo WR, Burggren WW. O_2 consumption and heart rate in developing zebrafish (Danio rerio): Influence of temperature and ambient O_2. American Journal of Physiology-Regulatory, Integrative and Comparative Physiology. 1999;**276**(2):R505-R513

[26] Lin L-Y, Cheng C-A, Liu S-T, Horng J-L. Investigation of ammonia-induced lethal toxicity toward ion regulation in zebrafish embryos. Comparative Biochemistry and Physiology Part C: Toxicology & Pharmacology.

2024;**276**:109807. DOI: 10.1016/j.cbpc.2023.109807

[27] Learmonth C, Carvalho AP. Acute and chronic toxicity of nitrate to early life stages of zebrafish - setting nitrate safety levels for zebrafish rearing. Zebrafish. 2015;**12**(4):305-311. DOI: 10.1089/zeb.2015.1098

[28] Villamizar N, Vera LM, Foulkes NS, Sánchez-Vázquez FJ. Effect of lighting conditions on zebrafish growth and development. Zebrafish. 2014;**11**(2):173-181. DOI: 10.1089/zeb.2013.0926

[29] Gatto E et al. Environmental enrichment decreases anxiety-like behavior in zebrafish larvae. Developmental Psychobiology. 2022;**64**(3):1-11. DOI: 10.1002/dev.22255

[30] Reiser S, Pohlmann DM, Koops U, Gröger JP, Focken U. Using gravel for environmental enrichment in salmonid hatcheries: The effect of gravel size during egg incubation, endogenous and first feeding in rainbow trout. Journal of Applied Ichthyology. 2019;**35**(2):465-472. DOI: 10.1111/jai.13884

[31] Kent ML, Buchner C, Barton C, Tanguay RL. Toxicity of chlorine to zebrafish embryos. Diseases of Aquatic Organisms. 2014;**107**(3):235-240. DOI: 10.3354/dao02683

[32] Winn AA, Prestia KA, Peneyra SM. Testing alternative surface disinfection agents for zebrafish (Danio rerio) embryos. Journal of the American Association for Laboratory Animal Science. 2021;**60**(5):510-518. DOI: 10.30802/AALAS-JAALAS-20-000164

[33] Varga ZM. Aquaculture, husbandry, and shipping at the Zebrafish International Resource Center. Methods in Cell Biology. 2016;**135**:509-534. Available from: http://zfin.org/ZDB-PUB-160725-11

[34] Félix L, Coimbra AM, Valentim AM, Antunes L. Review on the use of zebrafish embryos to study the effects of anesthetics during early development. Critical Reviews in Toxicology. 2019;**49**(4):357-370. DOI: 10.1080/10408444.2019.1617236

[35] Félix LM, Luzio A, Antunes L, Coimbra AM, Valentim AM. Malformations and mortality in zebrafish early stages associated with elevated caspase activity after 24 h exposure to MS-222. Toxicology and Applied Pharmacology. 2021;**412**:115385. DOI: 10.1016/j.taap.2020.115385

[36] Wisenden BD, Paulson DC, Orr M. Zebrafish embryos hatch early in response to chemical and mechanical indicators of predation risk, resulting in underdeveloped swimming ability of hatchling larvae. Biology Open. 2022;**11**(12):1-7. DOI: 10.1242/bio.059229

[37] Gonçalves I, Monteiro J. Colony management at zebrafish facilities. In: Zebrafish Research - An Ever-Expanding Experimental Model. London: IntechOpen; 2023. DOI: 10.5772/intechopen.1003707

[38] Gosálvez J, López-Fernández C, Hermoso A, Fernández JL, Kjelland ME. Sperm DNA fragmentation in zebrafish (Danio rerio) and its impact on fertility and embryo viability - implications for fisheries and aquaculture. Aquaculture. 2014;**433**:173-182. DOI: 10.1016/j.aquaculture.2014.05.036

[39] Konar ESM et al. Evaluation of viability, developmental competence, and apoptosis-related transcripts during in vivo post-ovulatory oocyte aging in zebrafish Danio rerio (Hamilton, 1822). Frontiers in Veterinary Science.

2024;**11**:1389070. DOI: 10.3389/fvets.2024.1389070

[40] OECD. Test No. 236: Fish Embryo Acute Toxicity (FET) Test. Test Guidelines 236. OECD Guidelines for the Testing of Chemicals, Section 2. Paris: OECD Publishing; 2025. DOI: 10.1787/9789264203709-en

[41] Beekhuijzen M et al. From cutting edge to guideline: A first step in harmonization of the zebrafish embryotoxicity test (ZET) by describing the most optimal test conditions and morphology scoring system. Reproductive Toxicology. 2015;**56**:64-76. DOI: 10.1016/j.reprotox.2015.06.050

[42] Pérez-Atehortúa M et al. Chorion in fish: Synthesis, functions and factors associated with its malformations. Aquaculture Reports. 2023;**30**:101590. DOI: 10.1016/j.aqrep.2023.101590

[43] Thomas JK, Janz DM. Embryo microinjection of selenomethionine reduces hatchability and modifies oxidant responsive gene expression in zebrafish. Scientific Reports. 2016;**6**:26520. DOI: 10.1038/srep26520

[44] Bruce AEE, Heisenberg C-P. Mechanisms of zebrafish epiboly: A current view. Current Topics in Developmental Biology. 2020;**136**:319-341. DOI: 10.1016/bs.ctdb.2019.07.001

[45] Kimmel CB, Ballard WW, Kimmel SR, Ullmann B, Schilling TF. Stages of embryonic development of the zebrafish. Developmental Dynamics. 1995;**203**(3):253-310. DOI: 10.1002/aja.1002030302

[46] Miller A et al. Temperature-dependence of early development of zebrafish and the consequences for laboratory use and animal welfare. bioRxiv. 2025:638062. DOI: 10.1101/2025.02.13.638062

[47] Santoso F et al. An overview of methods for cardiac rhythm detection in zebrafish. Biomedicine. 2020;**8**(9):329. DOI: 10.3390/biomedicines8090329

[48] Scholz S, Fischer S, Gündel U, Küster E, Luckenbach T, Voelker D. The zebrafish embryo model in environmental risk assessment - applications beyond acute toxicity testing. Environmental Science and Pollution Research. 2008;**15**(5):394-404. DOI: 10.1007/s11356-008-0018-z

[49] Segner H. Zebrafish (Danio rerio) as a model organism for investigating endocrine disruption. Comparative Biochemistry and Physiology Part C: Toxicology & Pharmacology. 2009;**149**(2):187-195. DOI: 10.1016/j.cbpc.2008.10.099

[50] Jang HS et al. Epigenetic dynamics shaping melanophore and iridophore cell fate in zebrafish. Genome Biology. 2021;**22**(1):282. DOI: 10.1186/s13059-021-02493-x

[51] Zindler F et al. Analysis of tail coiling activity of zebrafish (Danio rerio) embryos allows for the differentiation of neurotoxicants with different modes of action. Ecotoxicology and Environmental Safety. 2019;**186**:109754. DOI: 10.1016/j.ecoenv.2019.109754

[52] Sztal TE, Ruparelia AA, Williams C, Bryson-Richardson RJ. Using touch-evoked response and locomotion assays to assess muscle performance and function in zebrafish. Journal of Visualized Experiments. 2016;**116**:2016. DOI: 10.3791/54431

[53] Norton WHJ. Measuring larval zebrafish behavior: Locomotion, thigmotaxis, and startle. In: Zebrafish

Protocols for Neurobehavioral Research. Totowa, NJ: Humana Press; 2012. pp. 3-20. DOI: 10.1007/978-1-61779-597-8_1

[54] Sales CF et al. Thermal stress induces heat shock protein 70 and apoptosis during embryo development in a neotropical freshwater fish. Reproduction, Fertility, and Development. 2019;**31**(3):547-556. DOI: 10.1071/RD18217

[55] Jiang J, Chen Y, Yu R, Zhao X, Wang Q, Cai L. Pretilachlor has the potential to induce endocrine disruption, oxidative stress, apoptosis and immunotoxicity during zebrafish embryo development. Environmental Toxicology and Pharmacology. 2016;**42**:125-134. DOI: 10.1016/j.etap.2016.01.006

[56] Kirsten K et al. Acute and chronic stress differently alter the expression of cytokine and neuronal markers genes in zebrafish brain. Stress. 2021;**24**(1):107-112. DOI: 10.1080/10253890.2020.1724947

Chapter 7

Exploring Ultra-Fast Freezing Protocols for Zebrafish Sperm Cryopreservation

Sandra Martins, Inês Oliveira, Mariana Sampaio, Georgios Strimpakos, Guilherme Antunes, Joana Monteiro and Ana Catarina Certal

Abstract

Zebrafish sperm cryopreservation is becoming a necessary routine technique in research facilities, as genetically modified lines continue to expand within limited housing capacity. Enhancing this technology is therefore critical to reduce live fish stocks while preserving genetic pools. However, available protocols are often challenging to reproduce, and line recovery is often inconsistent and low yield. Here, we tested 23 variations of a sperm cryopreservation protocol based on ultra-fast freezing. Efficiency was evaluated via in vitro fertilisation (IVF) of wild-type oocytes with the cryopreserved sperm. Embryo survival was used as a readout of cryopreservation success. Four protocols demonstrated the highest embryo survival at 24 hours post-fertilisation (hpf); all shared the use of skim milk powder as an antioxidant and methanol as a cryoprotectant. Among these, the protocol incorporating Aquaboost® Spermcoat as the extender produced the best larval survival and lowest morbidity at 7 days post-fertilisation (dpf). This method was further optimised to increase genetic diversity and reduce variability in sperm quality by pooling sperm from five males, mixing it with extender, dividing the mixture into five cryovials containing freezing medium, and applying ultra-fast freezing. The resulting protocol, named Ultra-Fast Freezing (UFF), offers a more cost-effective and streamlined approach than other current alternatives. While technical proficiency remains crucial for reproducibility, this work contributes to the broader implementation of cryopreservation in zebrafish facilities as an important tool for genetic resource preservation, improved facility management, and alignment with the Replacement, Reduction, and Refinement (3Rs) principles.

Keywords: cryopreservation, zebrafish, sperm, IVF, ultra-fast freezing, 3Rs

1. Introduction

Every year, hundreds of new zebrafish transgenic, mutant, and edited lines are generated worldwide, while the animal holding capacity at fish facilities is often limited. This imposes difficult decisions on both researcher and facility teams on which fish lines to maintain and which to discontinue. Fortunately, nowadays it is possible to preserve the genomic pool of fish lines even when a living fish stock is no longer available, through gamete cryopreservation, namely sperm. The technique involves preserving sperm from fish of interest by exposing them to extreme cold, thereby maintaining the integrity of the cells. When a cryopreserved fish line is to be resuscitated, the sperm is thawed and used to in vitro fertilise freshly collected oocytes from a different line of interest (usually a wild-type line). Sperm cryopreservation not only prevents the loss of genomic pools but it also helps researchers save time, resources, and animals that would otherwise be needed to maintain generations with minimal inbreeding depression or to produce new lines formerly discontinued [1, 2]. In addition, through cryopreservation, two of the three principles of the 3Rs are met: reduction of the number of living fish stocks maintained at a given time and replacement of animal shipping with the shipping of sperm cells, when a fish line needs to be transferred to another facility [3].

Despite the existence of several published protocols for the cryopreservation of zebrafish oocytes [4, 5], ovarian tissue [5–8], and embryos [5, 8–10], none appears to have been fully effective, either due to the large volume of yolk or the more complex structure of these cells [4, 6]. The most successful protocols to date are for sperm cryopreservation [1, 11–13].

The major steps in sperm cryopreservation are sample collection, freezing, and storage, which must be well controlled to produce a viable cryopreserved sample [13]. Spermatozoa are immotile while inside the testis. Once the cells are expelled, they are activated by contact with water and gain motility. This motility phase lasts only a few seconds, gradually decreasing 30 seconds after activation and turning immotile and non-viable shortly after [11–16]. Therefore, sperm viability preservation depends greatly on the time window between collection and freezing, requiring it to be stored in a solution that replicates the testis environment while preserving all physiological and biological properties of the cell – the extender [12]. A freezing medium containing a mixture of the extender and one or more cryopreservation agents (CPAs) must therefore be added to the sample prior to freezing. CPAs dehydrate cells to prevent ice crystals from being formed and subsequent rupture of membranes [17], and the combination of the extender and CPAs allows for indefinite sperm preservation [18]. In addition, some protocols add an antioxidant substance to mitigate oxidative stress by neutralising reactive oxygen species (ROS), thereby preserving sperm function [19, 20]. ROS are often produced during the cryopreservation process, impairing sperm stability [21].

Most published protocols for zebrafish cryopreservation are laborious, error-prone, and poorly reproducible in different institutions [22]. For example, the most commonly used method relies on slow sperm freezing followed by storage in liquid nitrogen, which overlooks the risks of loss of motility and vitality, membrane damage, and increased ROS accumulation [8]. It has been reported that the use of a programmable freezer could improve this method, but the cost of such equipment may render it out of reach for many facilities or laboratories [23]. Another method, called vitrification, involves cell suspension in high concentrations of CPAs in combination with fast cooling rates [24–26]. If both CPAs and temperature rates are adequate, ice formation during cooling is prevented, which makes vitrification a fast-freezing method with several advantages over the slow-freezing method. Nevertheless, the use of high concentrations of CPAs may become toxic [12, 21, 27]. Thus, the development of more effective, reproducible, easier, and affordable methods will contribute to a more widespread implementation of cryopreservation of fish lines.

With the aim of developing a sperm cryopreservation protocol that would be easier to implement while also being less labour-intensive and more affordable, we propose here an optimised ultra-fast freezing method focused on temperature rates and concentration of cryoprotectants (CPAs). Different types, volumes, and combinations of known extenders and cryoprotectants were used. The survival rate at 24 hpf and general health status of the recovered offspring were used as indicators to evaluate cryopreservation success, as previously documented [11, 23, 28, 29]. The three best solution combinations were then selected for additional optimisation. We intended to create a sperm cryopreservation protocol that is easier to reproduce and that reduces the labour and costs involved.

2. Materials and methods

2.1 Zebrafish husbandry

Zebrafish wild-type and transgenic lines were used: AB, TU, Anju, Tg(tuba1a:GCaMP6s)ccu41; Tg(isl2b:Gal4), Tg(UAS:GFP); Tg(elavl3:GFF), Tg(UAS:mCherry); Tg(isl2b:Gal4), Tg(10xUAS:GCaMP6fEF05)ccu2; Tg(UAS:mCherry) (Champalimaud Foundation; Zebrafish International Resource Centre – ZIRC; Boston Children's Hospital). All fish were healthy adults, 3–18 months old.

Housing and husbandry of all animals were performed according to Martins et al. [30] study, with modifications according to Monteiro et al. [31].

To prepare the fish for gametes collection, a tank divider was inserted into each tank to separate males from females while remaining exposed to secreted pheromones, stimulating gamete production and maturation [32]. This physical separation lasted 2 weeks in phase 1 – initial assessment of the efficiency of 23 ultra-fast freezing sperm cryopreservation protocols – and 1 day in phase 2 – which focused on optimising the best-performing methods to enhance genetic diversity and reduce variability in sperm quality. The fish density in this set-up was 4 fish/L. Then, on the day of the gamete collection, fish were not fed. All embryos and larvae were housed in a dedicated incubator at 28°C with a photoperiod of 14-h light: 10-h dark and fed 1x/day with rotifers from 5 dpf onwards.

2.2 Sperm collection

Sperm collection was performed in the morning, between the first and third hours of light. The sperm collection procedure was adapted from the study by Draper and Moens [33]. Males were anaesthetised by immersion in 0.16 g/L MS-222, rinsed in PBS, and the excess solution from the ventral region briefly blotted out. They were then laid in a sponge holder soaked in PBS, ventral side up, under a stereoscope. The head of the fish was always carefully maintained out of the sponge to avoid any ocular lesions, and the urogenital region was dried with a Kimwipe® tissue. A 10-μL capillary (Blaubrand™) attached to a mouthpiece was used to reveal the urogenital pore from where up to 3 μL of sperm were collected with extremely gentle suction, while gently pressing the fish laterally with flat forceps (Fisher Scientific Cole-Parmer™). The male was then placed in a tank with system water and was kept under observation until complete recovery. Sperm quality was assessed qualitatively through bright white colour and dense concentration. Only good quality samples were frozen.

2.3 Determination of sperm concentration

Sperm concentration was estimated using a NanoDrop® microvolume spectrophotometer (Thermo Scientific, Wilmington, DE). The protocol was as previously established [34]. Briefly, 2–3 µL of sperm were suspended in 10 µL of extender, from which 1 µL loaded onto the NanoDrop, and absorbance was measured at 400 nm. The standard equation $Y = (3 \times 10^8) X - 3 \times 10^7$ with "X" representing the absorbance measured was used. Sperm concentration was assessed for a total of 28 samples.

2.4 Ultra-fast freezing

Different extenders and CPAs/anti-oxidants were combined into different test groups (A to H), each subdivided based on the volume of each reagent, the number of males sampled, the number of cryovials used, and the timing of vials' capping. Overall, 23 protocol variations (ID #1 to #23) were tested (**Table 1**). These combinations were designed based on previous protocols [11, 33–38].

Before sperm collection, the extender was aliquoted into 0.5 mL microcentrifuge tubes. The freezing medium was placed in 2 mL Nunc™ cryovials (or straws, for protocol 14) and incubated on ice. The ultra-fast freezing chamber was also set up: a styrofoam box and lid, with two cryovial racks inside, one on the top of the other, and liquid nitrogen covering the bottom rack only.

As soon as sperm was collected from one to five males (**Table 1**), it was added to the extender, mixed, and quickly transferred to the freezing medium in the cryovials. The cryovials were quickly transferred, either with or without a cap (depending on the protocol), into the freezing chamber. They were incubated for 15 minutes on the top support, to be in contact with the nitrogen vapour (not directly with liquid nitrogen). They were then capped (protocol-dependent) and submerged in liquid nitrogen. In P14, straws were capped and directly submerged in liquid nitrogen.

2.5 Efficiency assessment

2.5.1 Hormonal stimulation of ovulation

With the aim of maximising the quality and quantity of mature oocytes per female, we used hormonal stimulation to maximise oocyte maturation and ovulation before collection. Between the first and second hours of light, four to five selected females were placed on 1-L breeding tanks with 0.0015 mg/L of 17α,20β-dihydroxy-4-pregnen-3-one (β-DHP) diluted in system water (adapted from Walker and Streisinger [39]). Females were incubated for 4–6 hours.

2.5.2 Oocyte collection

Females were anaesthetised in 0.16 g/L MS-222, rinsed in PBS, the urogenital region dried with Kimtech® tissues, and each fish was held ventral side up with one dry hand, between the index and thumb. Using the same hand, gentle pressure was applied to the abdominal area in the rostrocaudal direction to expel the oocytes. The other hand was holding a polypropylene spatula next to the urogenital pore to immediately collect the oocytes. If no gametes came out upon gentle pressure, no additional pressure was applied. After the oocyte stripping procedure, the fish was transferred to a 1-L tank with fresh system water until complete recovery from anaesthesia. After recovery, all fish were returned to the recirculating holding system.

DOI: http://dx.doi.org/10.5772/intechopen.1013519

	Freezing medium										
Group	**Extender***	**Antioxidant***	**Cryoprotectant**	**Protocol**	**Extender (μL)**	**Freezing medium (μL)**	**# Sperm samples**	**Cryovials (n°)**	**Cap in N2 vapour**	**Survival rate at 24 hpf (mean % ± SD)**	**N**
A	Ginsburg	Skim milk powder	Methanol	1	3.3	20	1	1	no	9.37 ± 16.55	36
				2	3.3	20	1	1	yes	3.85 ± 7.27	8
				3	50	330	5	2	no	6.49 ± 7.75	12
				4	6.6	40	1	2	no	13.90 ± 14.09	15
				5	6	20	1	1	no	11.18 ± 8.58	5
B	Sperm coat	—	Methanol	6	3.3	20	1	1	no	0	3
				7	6.6	40	1	2	no	0	4
				8	55	100	5	4	no	0	8
				9	50	65	5	4	no	6.14 ± 7.42	8
C	HBSS	—	Methanol	10	3.3	20	1	1	no	0	6
				11	21.3	2	1	1	no	0	6
D	HBSS	—	DMA 10%	12	3.3	20	1	1	no	1.80 ± 2.34	6
			DMA 8%	13	3.3	20	1	1	no	0	6
				14	50	445	5	2◆	yes	2.17 ± 4.63	6
				15	10	15	1	2	no	1.37 ± 2.33	8
E	Sperm coat	Skim milk powder	Methanol	16	6.6	40	1	2	no	4.68 ± 4.32	4
				17	6	20	1	1	no	17.91 ± 17.65	11
				18	3.3	20	1	1	no	7.33 ± 12.17	10
				19	6.6	40	2	2	no	0.38 ± 0.86	12
F	Sperm coat	Catalase	Methanol	20	6	20	1	1	no	0	6
				21	3	20	1	1	no	0	8

		Freezing medium									
Group	**Extender***	**Antioxidant***	**Cryoprotectant**	**Protocol**	**Extender (μL)**	**Freezing medium (μL)**	**# Sperm samples**	**Cryovials (n°)**	**Cap in N2 vapour**	**Survival rate at 24 hpf (mean % ± SD)**	**N**
G	Ginsburg	Catalase	Methanol	22	3	20	1	1	no	0	3
H	Ginsburg	Skim milk powder	DMF	23	3.3	20	1	1	no	0	10

Protocols P1 to P23 are organised by groups A to H according to the extender and freezing medium used.
**Components of the Extender Medium. (◆) Straws used instead of cryotubes.*

Table 1.
Summary of protocols tested for wild-type (AB and TU) sperm ultra-fast freezing.

The quality of the oocytes was visually inspected immediately upon collection: samples with turbidity, white debris, dirt, or watery oocytes were discarded. Gametes that looked yellowish, dry, and stuck together were quickly but gently transferred and kept for 5–30 minutes into a petri dish with 75–100 μL of Aquaboost® Ovacoat, an extender preventing oocyte activation.

2.5.3 Thawing of cryopreserved sperm samples

The thawing protocol was adapted from Yang et al. [29] and performed in parallel with the oocyte collection. Briefly, for each cryovial to be thawed, 70 μL of sperm extender was incubated in a water bath at 33°C for 1 h. Oocytes were collected into Aquaboost® Ovacoat, then a frozen sperm cryovial was picked from the storage, the cap removed, and the vial tilted upside-down to discard any residual liquid nitrogen. The bottom half of the vial was quickly immersed in a 33°C water bath for 15 seconds, and the pre-heated extender was rapidly added and mixed by gently pipetting up and down two to three times using a 200 μL pipette tip previously cut to a diameter of approximately 1.5 mm. Sperm samples were incubated in the extender for a maximum of 30 minutes and used for in vitro fertilisation.

2.5.4 In vitro fertilisation (IVF) using cryopreserved sperm

The IVF protocol was optimised based on existing protocols from ZFIN [39], ZIRC [12], UCL, and Cryogenetics (https://www.cryogenetics.com/ivf/). Soon after sperm thawing, the Aquaboost® Ovacoat was totally removed from the oocytes using a 200-μL pipette tip. Sperm was immediately added, again using a 200-μL pipette with a cut tip, followed by 750 μL of system water. This mix was incubated for 5 minutes at room temperature (~25°C) without shaking. Petri dishes were then filled with system water and transferred to an incubator at 28°C. After 3–4 hours, water was replaced with embryo medium (E3).

2.5.5 Assessment of successful cryopreservation

Cryopreserved sperm samples were thawed and used to fertilise freshly collected oocytes. Progeny survival at 24 hpf and 7 dpf, as well as morbidity rate at 7 dpf, were used to assess the success of the ultra-fast freezing protocols, using the following formulae:

$$\text{Survival 24}\ hpf = (\text{live embryos/total number (embryos + non–fertilised oocytes)})^{*}100 \pm \text{standard deviation.} \tag{1}$$

$$\text{Morbidity 7}\ dpf = (\text{larvae with abnormal development/total number of larvae})^{*}100 \pm \text{standard deviation.} \tag{2}$$

$$\text{Survival 7}\ dpf = (\text{live larvae/total number larvae})^{*}100 \pm \text{standard deviation.} \tag{3}$$

2.6 Statistical analysis

Data were first tested for assumptions of normality of distribution and homogeneity of variances using Kolmogorov-Smirnov and Levene's test, respectively. The Kruskal-Wallis test followed by pairwise comparison was used for protocol comparison, and Pearson's correlation coefficient (r) between sperm concentration and fertilisation rates was calculated. Statistical analyses were performed at a significance level of 0.05, using Statistica™ software (Version 10.0, StatSoft Inc., Tulsa, Oklahoma, USA).

2.7 Reagents

Hormone 17α,20β-Dihydroxy-4-pregnen-3-one (P6285-5MG Sigma Aldrich) 100 nM;

Aquaboost® Ovacoat (Cryogenetics);

Aquaboost® Spermcoat (Cryogenetics);

Skim milk powder (70166-500G Fluka Analytical);

Ginsburg Ringer's Solution prepared as follows: in 450 ml sterile deionised water, dissolve NaCl 3.25 g, KCl 0.125 g, CaCl2•2 H2O 0.175 g, and NaHCO3 0.10 g. Bring final volume to 500 ml with sterile deionised water (all reagents from Sigma Aldrich);

Hanks' balanced salt solution at an osmolality of 300 mOsmol/kg abbreviated as HBSS300 prepared as follows: 0.137 M NaCl, 5.4 mM KCl, 1.3 mM $CaCl_2$, 1.0 mM $MgSO_4$, 0.25 mM Na_2HPO_4, 0.44 mM KH_2PO_4, 4.2 mM $NaHCO_3$, and 5.55 mM glucose, pH 7.2 (all reagents from Sigma Aldrich);

Catalase (C40-100MG Sigma Aldrich);

N, N-Dimethylformamide 99.8%, Extra Dry, AcroSeal™ (ACROS Organics) abbreviated as DMF;

N, N-dimethylacetamide 99.55 (38840-1 L-F Honeywell) abbreviated as DMA;

Methanol (UN1230 Fisher Scientific).

3. Results

3.1 Ultra-fast freezing optimisation

Ultra-fast freezing was attempted in a total of 201 samples across 23 protocol variations, organised into groups A to H, according to the extender and freezing medium. Frozen sperm samples were used to resuscitate the fish line through IVF, with freshly collected oocytes. Embryo survival at 24 hpf was used as a readout of the cryopreservation success.

In group A (Ginsburg + skim milk + methanol), the impact of absence (P1) or presence (P2) of the vial cap during the incubation step in nitrogen vapour was assessed, and upon IVF, there was a better survival rate in oocytes fertilised with sperm from uncapped vials (9.97% uncapped *vs.* 3.85% capped). Therefore, cryovials were left uncapped during the remainder of the tests.

Protocols in group E (sperm coat + skim milk + methanol) had the most variable survival rate, with significant differences between the lowest and the highest survival protocols, P19 ($0.38 \pm 0.90\%$) and P17 ($17.91 \pm 18.51\%$), respectively. Both protocols had ~28 µL of sperm solution per cryovial, whereas in P19, sperm from two males was collected, mixed, and divided into two aliquots; in P17, sperm from a single male was used to prepare a single sample, thus taking less time to complete the protocol, which could explain the results.

In group B (sperm coat + methanol), P9 was the only protocol that fertilised oocytes successfully. This may be related to the ratio of extender:cryoprotectant volume, as it was the only parameter that varied among all protocols in group B.

3.1.1 Effect of extender and freezing media

Protocols in groups C, F, G, and H did not yield viable embryos at 24 hpf. These comprised a freezing medium containing HBSS and methanol (P10–11), catalase (P20–22), or DMF (P23).

The protocols showing the higher survival rates at 24 hpf were P4 (group A) and P17 (group E), with P17 being the most successful (17.91 ± 18.51%). These protocols utilise the same cryoprotectant and antioxidant, namely skim milk and methanol, in different freezing media (**Table 1**). These results suggest that the combination of skim milk and methanol may play a crucial role in the success of the ultra-fast sperm freezing.

The beneficial effect of skim milk was further supported by the comparison of P6 (group B) and P18 (group E). Only P18, containing skim milk, produced live embryos. Similarly, the positive impact of methanol was further supported by comparing P1 (group A) with P23 (group H), where only P1, containing methanol, produced viable progeny at 24 hpf.

However, the efficiency of the cryoprotectant agents may still depend on the extender used. Supporting this, three out of four protocols combining HBSS with DMA (group D, P12–15) generated live embryos at 24 hpf (1.37 ± 2.49 to 2.17 ± 5.07%), whereas the combination of HBSS with methanol (group C, P10–11) failed to do so.

3.1.2 Survival rates and associated morbidity

The three most successful protocols (P1, P4, and P17) were selected for further analysis at 7 dpf, specifically examining the survival rate (**Figure 1**) and morbidity rate (**Figure 2**). P5 was not included in this analysis due to the lower N. Results did not reveal significant differences across the three protocols ($p > 0.05$), nevertheless being consistent with the survival rates obtained at 24 hpf (**Table 1**). P1 presented the lowest survival rate (9.37 ± 16.55%) at 24 hpf, which is consistent with the lowest survival rate at 7 dpf (87.25 ± 18.70%) and the highest morbidity rate at 7 dpf (2.88 ± 9.11%). P4 presented an intermediate result among the three protocols, showing a similar tendency at 24 hpf (13.90 ± 14.09), 7 dpf survival rate (89.29 ± 9.87%), and also for the 7 dpf morbidity rate (1.00 ± 2.09%). P17 showed the highest survival rate at 24 hpf (17.91 ± 17.65%) and at 7 dpf (96.97 ± 4.79%) and the lowest morbidity at 7 dpf (0.42 ± 1.18%). These data therefore suggest that the use of Sperm coat as an extender (P17) may be a better option than Ginsburg (P1).

3.2 Further optimisation of the best-performing ultra-fast freezing protocol

Based on the results above, protocol P17 was selected for a final optimisation step of an Ultra-Fast Freezing (UFF) protocol. The goal was to minimise sample variability in sperm quality and to maximise the genetic pool on each cryopreserved sample while improving fertilisation success and embryo viability rates. P17 was thus adapted

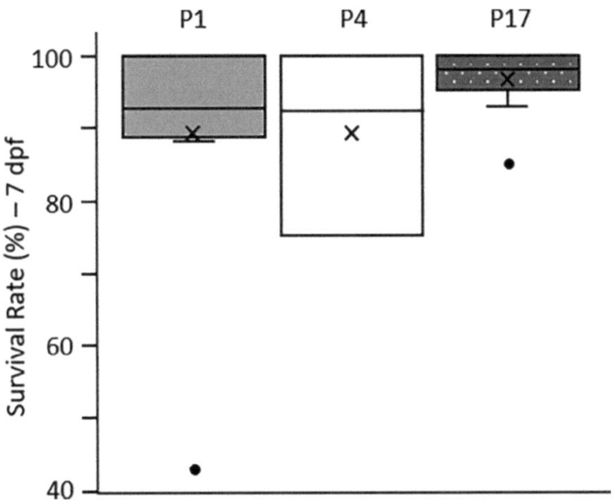

Figure 1.
Survival rate (%) at 7 dpf after IVF using sperm frozen using protocols P1, P4, and P17. Kruskal-Wallis test followed by pairwise comparison did not reveal significant differences ($p > 0.05$). Boxplot markers: minimum (lower whisker), quartile 1 (lower side of the box), median (line within the box), quartile 3 (upper side of the box), maximum (upper whisker), outliers (individual dots), and mean ("x" symbol).

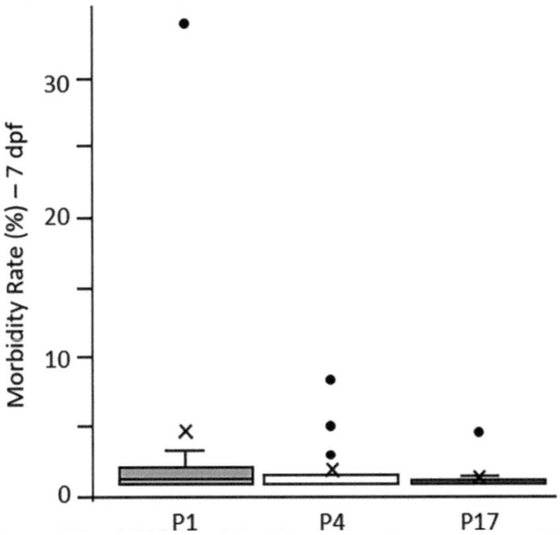

Figure 2.
Morbidity rate (%) at 7 dpf after IVF using sperm frozen using protocols P1, P4, and P17. Kruskal-Wallis test followed by pairwise comparison did not reveal significant differences ($p > 0.05$). Boxplot markers: minimum (lower whisker), quartile 1 (lower side of the box), median (line within the box), quartile 3 (upper side of the box), maximum (upper whisker), outliers (individual dots), and mean ("x" symbol).

to include sperm collection from five males, mixing all samples and then splitting again into five cryovials.

To evaluate the success of UFF (optimised version of P17), the survival rate at 24 hpf was determined and compared with the results from protocols P3, P9, P14, and P19, where each cryopreserved sample also included sperm from multiple fish (**Table 1, Figure 3**).

There was a higher survival rate for embryos generated with sperm from P3 than P19. UFF also led to better line recovery compared with P19. Interestingly, both UFF

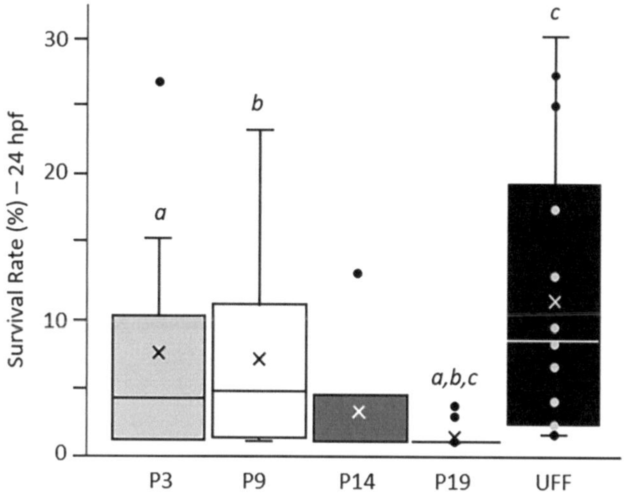

Figure 3.
Survival rate (%) at 24 hpf after IVF using cryopreserved sperm samples, each pooled from multiple males according to protocols P3, P9, P14, P19, and UFF. Results were analysed using the Kruskal-Wallis test followed by pairwise comparison. The survival rate for P19 was significantly lower than that for P3 (a: $p < 0.05$), P9 (b: $p < 0.05$), and UFF (c: $p < 0.005$). P14 was not significantly different from any of the other protocols. Boxplot markers: minimum (lower whisker), quartile 1 (lower side of the box), median (line within the box), quartile 3 (upper side of the box), maximum (upper whisker), outliers (individual dots), and mean ("x" symbol).

and P19 used the same extender, cryoprotectant and antioxidant. But, in P19, sperm was collected from only two males and was cryopreserved in a higher volume of freezing medium (thus at a lower concentration). In all other protocols, each sample included a mix of sperm from five males. Therefore, our results seem to support the initial hypothesis that mixing sperm from more males per sample benefits sample quality, as reflected by better percentages of line recovery.

3.3 Fertilisation success rate for different zebrafish lines

For the UFF protocol refinement, males from four transgenic zebrafish lines – tuba1a:GCaMP6s(ccu41Tg), elavl3:GFF(ccu5Tg) UAS:mCherry, isl2b:Gal4 10xUAS: GCaMP6fEF05(ccu2Tg), and UAS:mCherry – and one wild-type line – Anju – were used to understand if the genetic background could play a role. Survival rates were determined for each line separately (**Figure 4**). According to the results, cryopreservation success with UFF was similar among all transgenic lines, and between most of the transgenic lines and the wild type. The survival rate of Anju was only significantly lower than that of isl2b:Gal4 10xUAS:GCaMP6fEF05 ($p < 0.05$).

4. Discussion

The viability of cryopreserved cells depends on several factors, making it difficult to develop a successful and reproducible cryopreservation protocol [35, 36]. Here, we attempted to develop a less time-consuming and less-expensive method, which was named ultra-fast freezing (UFF). This protocol primarily differs from traditional cryopreservation and vitrification techniques in the freezing rate and the concentration of cryoprotectant agents.

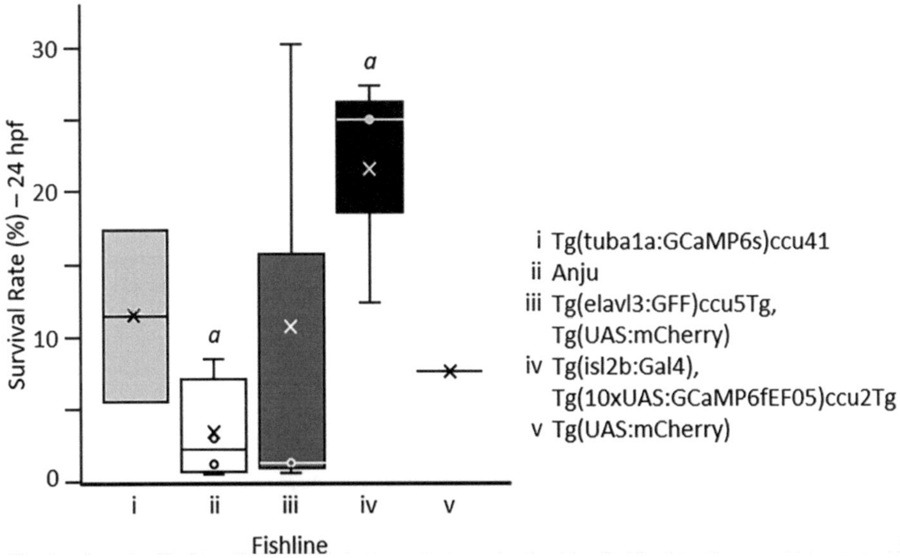

Figure 4.
Survival rate (%) at 24 hpf after IVF with sperm cryopreserved using the UFF protocol for each line (n = 13): (i) tuba1a:GCaMP6s(ccu41Tg). (ii) Anju. (iii) elavl3:GFF(ccu5Tg) UAS:mCherry. (iv) isl2b:Gal4 10xUAS: GCaMP6fEF05(ccu2Tg). (v) UAS:mCherry. Results were analysed using the Kruskal-Wallis test followed by pairwise comparison. The survival rate for Anju embryos (ii) was significantly lower than that for isl2b:Gal4 10xUAS:GCaMP6fEF05(ccu2Tg) (iv) (a: $p < 0.01$). No other significant differences were found ($p > 0.05$). Boxplot markers: minimum (lower whisker), quartile 1 (lower side of the box), median (line within the box), quartile 3 (upper side of the box), maximum (upper whisker), outliers (individual dots), and mean ("x" symbol).

The lack of dynamic development in fish cryopreservation methods can be partly attributed to the absence of alternative extenders that improve spermatozoa viability and to the frequent use of toxic cryoprotectants. When developing a protocol for fish sperm manipulation, it is crucial to consider physiological mechanisms, such as the absence of motility within the genital tract and the extremely short lifespan upon activation. In zebrafish, the natural spermatozoa lifespan is in the short minute range [11–16, 40], and this brief period of activity requires special attention to two important steps: the operational agility of the technician and the efficiency of the extender. For both extenders and cryoprotectants, intrinsic biophysical properties must also be considered, such as the cell's water and cryoprotectant permeability, osmotic tolerance limits [15], and intracellular ice nucleation [41, 42].

To reach an ultra-fast freezing (UFF) protocol with acceptable survival rates (that could sustain a reliable line recovery), 23 protocols were tested to evaluate which was most adequate for zebrafish sperm cryopreservation and if UFF could be a new optional choice for facilities.

4.1 Cryoprotectants, antioxidants, and extender as key players

Three extenders, two antioxidants, and three cryoprotectants were used in a total of eight different combinations of extender and freezing medium, resulting in 23 protocols. Ultimately, this led to the development of the UFF protocol, which combines different classes of cryoprotectants/antioxidants. Together with previous studies, this reinforces the idea that such a combination creates synergy, resulting in higher motility post-thaw, fewer sperm deformities, and less DNA damage [16, 38].

DOI: http://dx.doi.org/10.5772/intechopen.1013519

4.1.1 The extender

Aquaboost® Spermcoat was the extender showing the highest embryo survival/viability when combined with methanol and skim milk powder (P17, **Figures 2** and **3**). We believe the success of the combined use of these reagents may be due to a synergistic effect between the osmolality of the extender, the CPA concentration in methanol that drives dehydration, and the antioxidant properties of the skim milk powder. The less-performing Ginsburg's solution had been used in other cryopreservation protocols that were successful in post-thaw fertilisation, but in the context of whole-testis dissection or slow-freezing techniques [15, 33, 34].

From the three extenders tested, HBSS had the lowest survival rate (**Table 1**) despite its reported success in cryopreserving zebrafish sperm [36]. However, successful cryopreservation protocols using HBSS as an extender and methanol as a cryoprotectant were based on testis dissection, always with the addition of an antioxidant, such as skim milk powder [16, 35–37]. HBSS had also proved successful when combined with DMF as a cryoprotectant, along with egg yolk, BSA, and glycine as antioxidants, and stored in an electric freezer [38]. Our data also show that HBSS combined with DMA rendered the best results when using straws instead of cryovials, which is consistent with the literature [37].

We propose that in a future development, the effects of adding skim milk powder to HBSS and DMA are tested on survival rates when using the ultra-fast freezing method. Additionally, a combination of HBSS with the cryoprotectants methanol and DMF, alongside skim milk powder as an antioxidant, could also hold promise in this new method.

4.1.2 The antioxidant

The presence or absence of antioxidants can influence the survival rate of vertebrate embryos. According to the literature, the addition of antioxidants prevents membrane damage, reduces ROS, prevents DNA damage of sperm, and increases sperm motility, thus favouring the survival rate [18, 19, 21, 41, 43].

The mechanisms through which skim milk powder protects sperm from ROS and DNA damage are not well understood in zebrafish but seem to be related to the presence of casein micelles that help to reduce damage in the sperm membrane by preventing cholesterol and lipid loss [12]. Furthermore, the ability of casein micelles to reduce ROS, cell membrane, and DNA damage has been reported, as well as their effect on reducing sperm tail agglutination, which together improves the survival rate [12, 18, 44]. Our data show that adding skim milk powder has increased the survival rate in all protocols, highlighting the importance of skim milk powder as an antioxidant in ultra-fast freezing cryopreservation.

Catalase has been described as a first-line neutraliser of reactive species, by catalysing the reduction of H_2O_2 from peroxisomes to water and O_2 [43], and has been proven to reduce oxidative stress in zebrafish fresh sperm [24]. However, our protocols using catalase (P20–P22) were not successful in terms of embryo survival rate. Catalase, when combined with HBSS, showed good results when cooling sperm but not with freezing and thawing processes [44]. Moreover, catalase was only tested with HBSS. It would therefore be interesting to understand if combining catalase with HBSS and DMA would further increase the survival/success rate as happened here when combined with Ginsburg (P22) or Sperm coat (P20, P21).

In summary, our results suggest that for the ultra-fast freezing approach, skim milk is a better option than catalase as an antioxidant agent and that its addition to both extender and freezing medium benefits cell preservation.

4.1.3 The cryoprotectant

Methanol was the cryoprotectant that showed the most promising results, as expected from previous reports [8, 35, 38, 45]. However, it was only successful when combined with skim milk powder as an antioxidant (P1–5, P16–19, and UFF) (**Table 1**). Acting alone, methanol appears not to be sufficient to prevent damage to sperm cells: it may still protect against ice crystal formation through dehydration, but it does not seem capable of reducing oxidative stress and consequently cell damage [13].

DMA was successful when combined with HBSS, as previously reported [11]. In this combination without the addition of an antioxidant (P12–15), DMA behaved better than methanol (survival rates in **Table 1**). This result suggests that the addition of an antioxidant agent could further benefit survival results, and therefore, a new combination could still be explored. The cooling rate in our protocol could also need to be adjusted for DMA, as it has shown promising results with cooling rates around 20° C min^{-1} [36].

Although other studies have shown success in cryopreservation using DMF combined with other extenders and antioxidant agents [38], in this study, it was only tested in one protocol (P14) with skim milk and did not get viable embryos upon IVF (**Table 1**). Future experiments could potentially explore other useful combinations.

In addition, the efficiency of each cryoprotectant agent seems to be largely dependent on the extender to which it is combined. This is supported by three out of the four protocols that combined HBSS with DMA (group D, P12–15), generating viable embryos at 24 hpf (1.37 ± 2.49 to $2.17 \pm 5.07\%$), whereas the combination of HBSS with methanol (group C, P10–11) failed to do so, suggesting that DMA seems to be more effective than methanol when using HBSS.

In summary, our results suggest that methanol appears to be a more effective cryoprotectant than DMA or DMF.

4.2 Embryo survival rates as proxies of frozen sperm quality

Embryo survival rate upon IVF with frozen sperm was used as a readout of cryopreserved sperm quality after thawing. Our results showed a very high variability of embryo survival rates within each protocol and within each transgenic and wild-type lines (**Table 1** and **Figure 4**).

Sperm quality variability has been documented in zebrafish not only at the inter-line level but also at the intra-line level [46]. This variability is inherent to the housing and husbandry context since sperm quality can be affected by several parameters [47–48]. Such high variability poses a challenge to attaining replicable cryopreservation protocols.

Sperm quality assessment during this study was mainly based on visual inspection (qualitative), which can be prone to operator bias. However, all sperm samples were chosen following the same analysis criteria. Milkier sperm and larger volumes were selected as good-quality sperm samples, although water or urine could still be contaminating the sample, and thus inducing premature sperm activation and decreasing fertilisation success by diminishing motility after thawing [12].

Our results do establish a link between a higher sperm cell concentration and a higher embryo survival rate (**Figure 1**), although a minimal value of 10^6 cells/ml of sperm concentration appears to be necessary for an IVF success of ≥70%, as previously mentioned in the literature [49]. Higher sperm cell concentrations should in principle require higher volumes of extender and freezing solution to ensure all cells are preserved and protected from damage.

Initially, sperm concentration was assessed rather than sperm motility. We now feel that this was not the best approach as sperm motility seems to be far more important and directly correlated with the fertilisation success, and consequently embryo survival (higher motility, higher fertilisation success rate) [37, 50]. However, even ensuring good sperm concentrations and motility upon collection may not be enough to attain successful cryopreservation. Sperm cells seem to display a variable ability to endure freezing and recover the hypermotile state observed before cryopreservation. We hypothesise that this may be related not only to the cryopreservation protocol per se but also to the animal's intrinsic physiological conditions impacted by factors, such as health, feeding, age, and holding density, as described by others [48–50]. Even though most of the fish in our facility have very consistent and similar housing and husbandry conditions, a certain degree of individual variation cannot be ruled out, such as the amount of food intake, individual stress, the number of fish in each tank, and genetic background.

Fish pathogens can also impact gamete quality, such as *Pseudoloma neurophilia* that can infect gonads and brain areas involved in hormone synthesis, affecting gametogenesis and reproduction [51]. However, since our fish facility runs a comprehensive health programme and is specific pathogen-free (SPF) for the most relevant pathogens, we can exclude this as a variable parameter.

4.3 Recovering fish lines from cryopreserved sperm

Viable progeny resulting from IVF with frozen sperm was assessed for P1, P4, and P17, combining two different parameters – embryo survival rate at 24 hpf and embryo developmental malformation rate at 7 dpf. P1 and P4 have used single-male sperm collection and the same extender and freezing combination. However, P4 used double the volume of the extender and freezing solution, and the final solution was divided into two vials. Interestingly, P4 produced similar survival rates compared to P1 (89.29 ± 9.87% and 87.25 ± 18.70%, respectively), but P4 showed less variation (**Figure 1**). Regarding developmental malformation rates, P4 displayed a lower mean malformation rate at 7 dpf as compared to P1 (1.00 ± 2.09% *vs.* 3.71 ± 9.83%, respectively). This may reflect a more adequate freezing medium volume/sperm concentration ratio in P4 than in P1, further suggesting that extender, cryoprotectant, and antioxidant agents could have a better contact with cells and contribute to decreasing DNA and cell damage and, consequently, malformations in larvae [28].

4.4 Towards an optimal ultra-fast freezing (UFF) protocol

The further optimised P17, subsequently named as UFF protocol, presented a significantly higher embryo survival when compared to all other protocols using pooled sperm samples, e.g. P19 (0.38 ± 0.86%; $p < 0.01$).

The UFF protocol was also tested with transgenic lines, in contrast to the other protocols, which were only tested in wild-type lines. Although a recent study showed no significant differences in the reproductive success between wild type, mutants, and several transgenic lines [52], the influence of the genetic modifications on the reproductive physiology and gamete performance remains poorly studied [53, 54].

The lower survival rates in pooled samples (UFF *vs.* P17) may be related to the time span between sperm collection and freezing [55]. However, multiple fish collections also contribute to genetic heterogeneity and reduce the quality variability (when compared to single collections).

4.5 The importance of thawing and IVF

The specifics of the sperm thawing and the IVF protocol can greatly contribute to the variability in the embryo survival results.

An incorrect or fluctuating temperature of the water bath could impair fertilisation and thus embryo survival by altering the thawing temperature ramp. A difference as small as 5°C/min in the cooling rate has been described as impacting post-thaw cell survivability [12].

The oocyte collection and selection procedures in IVF could also dictate the fate of the progeny. Previous studies have shown that the oocyte-to-sperm ratio can determine the fertilisation success and that pooled oocyte clutches for each sperm sample can be ideal [38]. In this study, oocyte-to-sperm ratios were not rigorously controlled, which could account for the sample-to-sample variation. In addition, oocyte handling may also damage the cells' membranes and reduce fertilisation or lead to early oocyte activation, diminishing the likelihood of oocyte fertilisation [49]. Finally, cryoprotectants used for sperm freezing can also be toxic for oocytes, decreasing viability and, consequently, the likelihood of successful fertilisation [56].

The sperm cells, when undergoing cryopreservation, may suffer DNA and membrane damage, which may result in non-viable sperm cells or damaged sperm cells, giving rise to malformed embryos [24, 38, 57]. Healthy oocytes have internal DNA damage repair mechanisms that are capable to a certain extent of reverting some of this damage, increasing the chances of successful fertilisation and embryo survival [58, 59].

5. Conclusions

Protocols for zebrafish sperm cryopreservation are still scarce, and globally, fish facilities have poor investment in further development and refinement. During this study, we found it particularly difficult to find adequate extenders, CPAs, and antioxidants, and to determine the adequate ratios of reagents and cells for achieving a successful ultra-fast freezing methodology. Sample variability, together with error-prone manual procedures, was also challenging.

Although further optimisation is still needed, we believe to have contributed to the development of a new ultra-fast freezing protocol (UFF) showing promising results in frozen line recovery, which, in our opinion, is easier to execute than other available protocols and does not require expensive equipment.

Acknowledgements

The authors thank Zoltan Varga, Joy Murphy, Katy Murray, and Patrícia Diogo for insightful discussion. This work was funded by the Champalimaud Foundation with support from the research infrastructure Congento, co-financed by Lisboa Regional Operational Programme (Lisboa2020), under the PORTUGAL 2020 Partnership Agreement, through the European Regional Development Fund (ERDF) and Fundação para a Ciência e Tecnologia (Portugal) under the project LISBOA-01-0145-FEDER-022170.

Conflict of interest

The authors declare no conflict of interest.

Author details

Sandra Martins[1,2,3], Inês Oliveira[1], Mariana Sampaio[1], Georgios Strimpakos[4], Guilherme Antunes[1,5], Joana Monteiro[1] and Ana Catarina Certal[1*]

1 Champalimaud Foundation, Lisbon, Portugal

2 MARE Guia Marine Laboratory, Marine and Environmental Sciences Centre, University of Lisbon, Cascais, Portugal

3 Comparative Molecular and Integrative Biology Lab, Centre of Marine Sciences of the Algarve, University of Algarve, Faro, Portugal

4 Institute of Biochemistry and Cell Biology, CNR, Rome, Italy

5 ISPA - University Institute, Lisbon, Portugal

*Address all correspondence to: ana.certal@research.fchampalimaud.org

References

[1] Carmichael C, Westerfield M, Varga ZM. Cryopreservation and in vitro fertilization at the zebrafish international resource center. Methods in Molecular Biology (Clifton, N.J.). 2009;**546**:45-65. DOI: 10.1007/978-1-60327-977-2_4

[2] Hart-Johnson S, Mankelow K. Archiving genetically altered animals: A review of cryopreservation and recovery methods for genome edited animals. Laboratory Animals. 2021;**56**(1):26-34. DOI: 10.1177/00236772211007306

[3] Sneddon LU, Halsey LG, Bury NR. Considering aspects of the 3Rs principles within experimental animal biology. The Journal of Experimental Biology. 2017; **220**(17):3007-3016. DOI: 10.1242/jeb.147058

[4] Guan M, Rawson DM, Zhang T. Cryopreservation of zebrafish (Danio rerio) oocytes by vitrification. Cryo Letters. 2010;**31**(3):230-238

[5] Martínez-Páramo S, Horváth Á, Labbé C, Zhang T, Robles V, Herráez P, et al. Cryobanking of aquatic species. Aquaculture. 2017;**472**:156-177. DOI: 10.1016/j.aquaculture.2016.05.042

[6] Anil S. Development of in-vitro culture and cryopreservation protocol for zebrafish (Danio rerio) ovarian tissue fragments (thesis). United Kingdom: University of Bedfordshire; 2013

[7] Godoy L, Streit DP, Zampolla T, Bös-Mikich A, Zhang T. A study on the vitrification of stage III zebrafish (Danio rerio) ovarian follicles. Cryobiology. 2013;**67**(3):347-354. DOI: 10.1016/j.cryobiol.2013.10.002

[8] Marques LS, Fossati AAN, Rodrigues RB, Da Rosa HT, Izaguirry AP, Ramalho JB, et al. Slow freezing versus vitrification for the cryopreservation of zebrafish (Danio rerio) ovarian tissue. Scientific Reports. 2019;**9**(1):1-11. DOI: 10.1038/s41598-019-51696-7

[9] Khosla K, Wang Y, Hagedorn M, Qin Z, Bischof J. Gold nanorod induced warming of embryos from the cryogenic state enhances viability. ACS Nano. 2017;**11**(8):7869-7878. DOI: 10.1021/acsnano.7b02216

[10] Khosla K, Kangas J, Liu Y, Zhan L, Daly J, Hagedorn M, et al. Cryopreservation and laser nanowarming of zebrafish embryos followed by hatching and spawning. Advanced Biosystems. 2020;**4**(11): 2000138. DOI: 10.1002/adbi.202000138

[11] Berghmans S, Morris JP, Kanki JP, Look AT. Zebrafish sperm cryopreservation. Methods in Cell Biology. 2004;**77**:645-658. DOI: 10.1016/s0091-679x(04)77034-x

[12] Matthews JL, Murphy JM, Carmichael C, Yang H, Tiersch T, Westerfield M, et al. Changes to extender, cryoprotective medium, and in vitro fertilization improve zebrafish sperm cryopreservation. Zebrafish. 2018;**15**(3):279-290. DOI: 10.1089/zeb.2017.1521

[13] Matthews JL, Varga ZM. Cryopreservation of pooled sperm samples. Methods in Molecular Biology. 2021;**2218**:99-115. DOI: 10.1007/978-1-0716-0970-5_9

[14] Ingermann RL, Schultz CLF, Kanuga MK, Wilson-Leedy JG. Metabolism of motile zebrafish sperm. Comparative Biochemistry and Physiology Part A: Molecular &

Integrative Physiology. 2011;**158**(4): 461-467. DOI: 10.1016/j.cbpa.2010. 12.008

[15] Cheng Y, Franěk R, Rodina M, Xin M, Cosson J, Zhang S, et al. Optimization of sperm management and fertilization in zebrafish (Danio rerio (Hamilton)). Animals. 2021;**11**(6):1558. DOI: 10.3390/ani11061558

[16] Beckham J, Alam F, Omojola V, Scherr T, Guitreau A, Melvin A, et al. A microfluidic device for motility and osmolality analysis of zebrafish sperm. Biomedical Microdevices. 2018;**20**(3):67. DOI: 10.1007/s10544-018-0308-2

[17] Lahnsteiner F. The effect of internal and external cryoprotectants on zebrafish (Danio rerio) embryos. Theriogenology. 2008;**69**(3):384-396. DOI: 10.1016/j.theriogenology.2007. 10.007

[18] Rodrigues RB, Uczay M, Brito VB, Nunes Fossati AA, Godoy AC, Moura DJ, et al. Skim milk powder used as a non-permeable cryoprotectant reduces oxidative and DNA damage in cryopreserved zebrafish sperm. Cryobiology. 2020;**97**:76-84. DOI: 10.1016/j.cryobiol.2020.10.005

[19] Nezhad NC, Vahabzadeh Z, Allahveisie A, Rahmani K, Raoofi A, Rezaie MJ, et al. The effect of L-carnitine and coenzyme Q10 on the sperm motility, DNA fragmentation, chromatin structure and oxygen free radicals during, before and after freezing in oligospermia men. Urology Journal. 2021;**18**(03):330-336. DOI: 10.22037/uj. v18i03.6400

[20] Hungerford AJ, Bakos HW, Aitken RJ. Analysis of sperm separation protocols for isolating cryopreserved human spermatozoa. Reproduction & Fertility. 2023;**4**(2):e220133. DOI: 10.1530/RAF-22-0133

[21] Hagedorn M, McCarthy M, Carter VL, Meyers SA. Oxidative stress in zebrafish (Danio rerio) sperm. PLoS One. 2012;**7**(6):e39397. DOI: 10.1371/ journal.pone.0039397

[22] Caetano Da Silva C, Kollár T, Csenki-Bakos Z, Fernandes BHV, Horváth Á, Carvalho LR. An easy method for cryopreservation of zebrafish (Danio rerio) sperm. Zebrafish. 2019;**16**(3): 321-323. DOI: 10.1089/zeb. 2018.1696

[23] Diogo P, Martins G, Quinzico I, Nogueira R, Gavaia PJ, Cabrita E. Electric ultrafreezer (−150°C) as an alternative for zebrafish sperm cryopreservation and storage. Fish Physiology and Biochemistry. 2018; **44**(6):1443-1455. DOI: 10.1007/ s10695-018-0500-6

[24] Marques LS, Bos-Mikich A, Godoy LC, Silva LA, Maschio D, Zhang T, et al. Viability of zebrafish (Danio rerio) ovarian follicles after vitrification in a metal container. Cryobiology. 2015;**71**(3):367-373. DOI: 10.1016/j.cryobiol.2015.09.004

[25] Mphaphathi M, Ledwaba M, Thema M. The current status of semen and oocytes cryopreservation. In: Quain M, Koprowski R, editors. Cryopreservation - Applications and Challenges. Biomedical Engineering: IntechOpen; 2023. ch 2. DOI: 10.5772/intechopen. 107404

[26] Xin M, Abdul M, Dzyuba B, Cuevas-Uribe R, Shaliutina-Kolešová A, Linhart O. Progress and challenges of fish sperm vitrification: A mini review. Theriogenology. 2017;**98**:16-22. DOI: 10.1016/j.theriogenology.2017. 04.043

[27] Best BP. Cryoprotectant toxicity: Facts, issues, and questions.

Rejuvenation Research. 2015;**18**(5): 422-436. DOI: 10.1089/rej.2014.1656

[28] Lestari S, None Abinawanto A, Bowolaksono RG, Kristanto AH. The percentage of embryo viability after 48h sperm cryopreservation: Effect of various natural cryoprotectant. IOP Conference Series Earth and Environmental Science. 2020;**441**(1): 012070–0. DOI: 10.1088/1755-1315/441/1/012070

[29] Yang H, Hu E, Tiersch T, Carmichael C, Matthews J, Varga ZM. Temporal and concentration effects of methanol on cryopreservation of zebrafish (Danio rerio) sperm. Zebrafish. 2020;**17**(4):233-242. DOI: 10.1089/zeb.2019.1849

[30] Martins S, Monteiro JF, Vito M, Weintraub D, Almeida J, Certal AC. Toward an integrated zebrafish health management program supporting cancer and neuroscience research. Zebrafish. 2016;**13**(S1):S-47-S-55. DOI: 10.1089/zeb.2015.1198

[31] Monteiro J, Martins S, Farias M, Costa T, Certal A. The impact of two different cold-extruded feeds and feeding regimens on zebrafish survival, growth and reproductive performance. Journal of Developmental Biology. 2018;**6**(3):15. DOI: 10.3390/jdb6030015

[32] Wasden MB, Roberts RL, DeLaurier A. Optimizing sperm collection procedures in zebrafish. Journal of the South Carolina Academy of Science. 2017;**15**(2):7

[33] Draper BW, Moens CB. A high-throughput method for zebrafish sperm cryopreservation and in vitro fertilization. Journal of Visualized Experiments. 2009;**29**:1395. DOI: 10.3791/1395

[34] Harvey B, Norman Kelley R, Ashwood-Smith MJ. Cryopreservation of zebra fish spermatozoa using methanol. Canadian Journal of Zoology. 1982; **60**(8):1867-1870. DOI: 10.1139/z82-242

[35] Yang H, Daly J, Carmichael C, Matthews J, Varga ZM, Tiersch T. A procedure-spanning analysis of plasma membrane integrity for assessment of cell viability in sperm cryopreservation of zebrafish Danio rerio. Zebrafish. 2016; **13**(2):144-151. DOI: 10.1089/zeb.2015.1176

[36] Butts IAE, Mokdad A, Trippel EA, Pitcher TE. Development of a sperm cryopreservation protocol for redside dace: Implications for genome resource banking. Transactions of the American Fisheries Society. 2013;**142**(3):671-680. DOI: 10.1080/00028487.2012.756433

[37] Yang H, Carmichael C, Varga Z, Tiersch TR. Development of a simplified and standardized protocol with potential for high-throughput for sperm cryopreservation in zebrafish Danio rerio. Theriogenology. 2007;**68**(2): 128-136. DOI: 10.1016/j.theriogenology.2007.02.015

[38] Diogo P, Martins G, Nogueira R, Marreiros A, Gavaia PJ, Cabrita E. Cryoprotectants synergy improve zebrafish sperm cryopreservation and offspring skeletogenesis. Cryobiology. 2019;**91**:115-127. DOI: 10.1016/j.cryobiol.2019.10.001

[39] Walker C, Streisinger G. Embryo production by in vitro fertilization. In: Westerfield M, editor. Chapter 2 – Breeding. The Zebrafish Book - A Guide for the Laboratory Use of Zebrafish (Danio rerio). 5th ed. Eugene: University of Oregon Press; 2007

[40] Blackburn H, Torres L, Liu Y, Tiersch TR. The need for a framework

addressing the temporal aspects of fish sperm motility leading to community-level standardization. Zebrafish. 2022; **19**(4):119-130. DOI: 10.1089/zeb.2022.0006

[41] Hungerford A, Bakos HW, Aitken RJ. Sperm cryopreservation: Current status and future developments. Reproduction, Fertility and Development. 2022;**35**(3):265-281 DOI: 10.1071/RD22219

[42] Hagedorn M, Peterson A, Mazur P, Kleinhans FW. High ice nucleation temperature of zebrafish embryos: Slow-freezing is not an option. Cryobiology. 2004;**49**(2):181-189. DOI: 10.1016/j.cryobiol.2004.07.001

[43] Sandoval-Vargas L, Jiménez MS, González JR, Villalobos EF, Cabrita E, Isler IV. Oxidative stress and use of antioxidants in fish semen cryopreservation. Reviews in Aquaculture. 2020;**13**(1):365-387. DOI: 10.1111/raq.12479

[44] Yang H, Hu E, Matthews JL, Varga ZM, Tiersch TR. Is catalase an effective additive to alleviate oxidative stress during cryopreservation of zebrafish sperm at the repository level? Cryobiology. 2022;**104**:70-78. DOI: 10.1016/j.cryobiol.2021.10.003

[45] Depincé A, Gabory A, Dziewulska K, Le Bail PY, Jammes H, Labbé C. DNA methylation stability in fish spermatozoa upon external constraint: Impact of fish hormonal stimulation and sperm cryopreservation. Molecular Reproduction and Development. 2020;**87**(1):124-134. DOI: 10.1002/mrd.23297

[46] Paull GC, Van Look KJ, Santos EM, Filby AL, Gray DM, Nash JP, et al. Variability in measures of reproductive success in laboratory-kept colonies of zebrafish and implications for studies addressing population-level effects of environmental chemicals. Aquatic Toxicology. 2008;**87**(2):115-126. DOI: 10.1016/j.aquatox.2008.01.008

[47] Kowalski RK, Cejko BI. Sperm quality in fish: Determinants and affecting factors. Theriogenology. 2019; **135**:94-108. DOI: 10.1016/j.theriogenology.2019.06.009

[48] Bobe J, Labbé C. Egg and sperm quality in fish. General and Comparative Endocrinology. 2010;**165**(3):535-548. DOI: 10.1016/j.ygcen.2009.02.011

[49] Hagedorn M, Carter VL. Zebrafish reproduction: Revisiting in vitro fertilization to increase sperm cryopreservation success. PLoS One. 2011;**6**(6):e21059. DOI: 10.1371/journal.pone.0021059

[50] Guo W, Xie B, Xiong S, Liang X, Gui JF, Mei J. miR-34a regulates sperm motility in zebrafish. International Journal of Molecular Sciences. 2017; **18**(12):2676. DOI: 10.3390/ijms18122676

[51] Spagnoli ST, Xue L, Murray KN, Chow F, Kent ML. Pseudoloma neurophilia: A retrospective and descriptive study of nervous system and muscle infections, with new implications for pathogenesis and behavioral phenotypes. Zebrafish. 2015;**12**(2): 189-201. DOI: 10.1089/zeb.2014.1055

[52] De Oliveira J, Chadili E, Turies C, Brion F, Cousin X, Hinfray N. A comparison of behavioral and reproductive parameters between wild-type, transgenic and mutant zebrafish: Could they all be considered the same "zebrafish" for reglementary assays on endocrine disruption? Comparative Biochemistry and Physiology Part C: Toxicology & Pharmacology. 2021;**239**: 108879. DOI: 10.1016/j.cbpc.2020.108879

[53] Mustafa A, Roman E, Winberg S. Boldness in male and female zebrafish (Danio rerio) is dependent on strain and test. Frontiers in Behavioral Neuroscience. 2019;**13**:248. DOI: 10.3389/fnbeh.2019.00248

[54] Egan RJ, Bergner CL, Hart PC, Cachat JM, Canavello PR, Elegante MF, et al. Understanding behavioral and physiological phenotypes of stress and anxiety in zebrafish. Behavioural Brain Research. 2009;**205**(1):38-44. DOI: 10.1016/j.bbr.2009.06.022

[55] Hagedorn M, Ricker J, McCarthy M, Meyers SA, Tiersch TR, Varga Z, et al. Biophysics of zebrafish (Danio rerio) sperm. Cryobiology. 2009;**58**(1):12-19. DOI: 10.1016/j.cryobiol.2008.09.013

[56] Plachinta M, Zhang T, Rawson DM. Studies on cryoprotectant toxicity to zebrafish (Danio rerio) oocytes. Cryo Letters. 2004;**25**(6):415-424

[57] Riesco MF, Robles V. Correction: Cryopreservation causes genetic and epigenetic changes in zebrafish genital ridges. PLoS One. 2014;**9**(1):e67614. DOI: 10.1371/annotation/3265139d-64c7-4c4c-83d3-1e139031e7df

[58] Ahmadi A, Ng SC. Fertilizing ability of DNA-damaged spermatozoa. The Journal of Experimental Zoology. 1999; **284**(6):696-704

[59] Cabrita E, Robles V, Rebordinos L, Sarasquete C, Herráez MP. Evaluation of DNA damage in rainbow trout (Oncorhynchus mykiss) and gilthead sea bream (Sparus aurata) cryopreserved sperm. Cryobiology. 2005;**50**(2): 144-153. DOI: 10.1016/j.cryobiol.2004.12.003